OPTICAL CONTROL of MICROWAVE DEVICES

The Artech House Microwave Library

Analysis, Design, and Applications of Fin Lines, Bharathi Bhat and Shiban K. Koul
E-Plane Integrated Circuits, P. Bhartia and P. Pramanick, eds.
Filters with Helical and Folded Helical Resonators, Peter Vizmuller
GaAs MESFET Circuit Design, Robert A. Soares, ed.
Gallium Arsenide Processing Techniques, Ralph Williams
Handbook of Microwave Integrated Circuits, Reinmut K. Hoffmann
Handbook for the Mechanical Tolerancing of Waveguide Components, W.B.W. Alison
Handbook of Microwave Testing, Thomas S. Laverghetta
High Power Microwave Sources, Victor Granatstien and Igor Alexeff, eds.
Introduction to Microwaves, Fred E. Gardiol
LOSLIN: Lossy Line Calculation Software and User's Manual, Fred. E. Gardiol
Lossy Transmission Lines, Fred E. Gardiol
Materials Handbook for Hybrid Microelectronics, J.A. King, ed.
Microstrip Antenna Design, K.C. Gupta and A. Benalla, eds.
Microstrip Lines and Slotlines, K.C. Gupta, R. Garg, and I.J. Bahl
Microwave Engineer's Handbook: 2 volume set, Theodore Saad, ed.
Microwave Filters, Impedance Matching Networks, and Coupling Structures, G.L. Matthaei, L. Young and E.M.T. Jones
Microwave Integrated Circuits, Jeffrey Frey and Kul Bhasin, eds.
Microwaves Made Simple: Principles and Applications, Stephen W. Cheung, Frederick H. Levien, et al.
Microwave and Millimeter Wave Heterostructure Transistors and Applications, F. Ali, ed.
Microwave Mixers, Stephen A. Maas
Microwave Transition Design, Jamal S. Izadian and Shahin M. Izadian
Microwave Transmission Line Filters, J.A.G. Malherbe
Microwave Transmission Line Couplers, J.A.G. Malherbe
Microwave Tubes, A.S. Gilmour, Jr.
MMIC Design: GaAs FETs and HEMTs, Peter H. Ladbrooke
Modern Spectrum Analyzer Theory and Applications, Morris Engelson
Monolithic Microwave Integrated Circuits: Technology and Design, Ravender Goyal, et al.
Nonlinear Microwave Circuits, Stephen A. Maas
Terrestrial Digital Microwave Communications, Ferdo Ivanek, et al.

OPTICAL CONTROL of MICROWAVE DEVICES

RAINEE SIMONS
NASA Lewis Research Center
Cleveland, Ohio

ARTECH HOUSE
BOSTON • LONDON

Library of Congress Cataloging-in-Publication Data

Simons, Rainee, N., 1949-
  Optical control of microwave devices / Rainee N. Simons.
     p.   cm.
  Includes bibliographical references .
  ISBN 0-89006-313-3
  1. Integrated optics. 2. Microwave integrated circuits.
  I. Title.
  TA1660.S55 1990   89-49101
  621.381'32--dc20   CIP

British Library Cataloguing in Publication Data

Simons, Rainee N., 1949-
  Optical control of microwave devices
  1. Microwave equipment. Circuits
  I. Title
  621.38132

  ISBN 0-89006-313-3

Copyright © 1990

ARTECH HOUSE
685 Canton Street
Norwood, MA 02062

All rights reserved. Printed and bound in the United States of America. No part of this publication may be reproduced or utilized in any form or by any means, electronic or mechanical, including photocopying, recording, or by any information storage and retrieval system, without permission in writing from the publisher.

International Standard Book Number: 0-89006-313-3
Library of Congress Catalog Card Number: 89-49101

  10   9   8   7   6   5   4   3   2   1

*To
Joy,
Renita, and
Rona*

Contents

Foreword			xv
Preface			xvii
Chapter 1	INTRODUCTION		1
1.1	Advantages of Optical Microwave Monolithic Integrated Circuits		2
	1.1.1	Performance	2
	1.1.2	Design	3
	1.1.3	Manufacturing	3
1.2	Typical Optical Microwave Monolithic Integrated Circuits		3
	1.2.1	Laser Diode and Field-Effect Transistor	3
	1.2.2	Laser Diode and Gunn Diode	3
	1.2.3	Laser Diode and Heterojunction Bipolar Transistor	4
	1.2.4	Laser Diode and an Optical Intensity Modulator	4
	1.2.5	Metal-Semiconductor-Metal Photodetector and Field-Effect Transistor	4
	1.2.6	PIN Photodetector and Field-Effect Transistor	4
	1.2.7	Laser Diode, Photodetector, and Field-Effect Transistor	5
1.3	Typical Applications		5
	1.3.1	Beam Steering in Phased Arrays	5
	1.3.2	Reference Frequency Distribution in Phased Arrays	5
	1.3.3	Remoting of Antennas	6
	1.3.4	Signal Processing and Electronic Warfare	6
	1.3.5	Missile Guidance	7
	1.3.6	Cable Television	7
	1.3.7	Cellular Telephone	7
	1.3.8	Instrumentation	7

1.4	Organization of the Book	8
	References	9

Chapter 2 LASER DIODES — 13

- 2.1 Introduction — 13
- 2.2 The Concept of Stimulated Emission — 14
- 2.3 The Double Heterostructure Laser Diodes — 15
  - 2.3.1 Energy Band Structure — 16
  - 2.3.2 Threshold Current Density — 17
- 2.4 Waveguiding Principles and Intensity Pattern of a Double Heterostructure Laser Diode — 20
  - 2.4.1 Even TE Modes — 22
  - 2.4.2 Odd TE Modes — 25
  - 2.4.3 TM Modes — 25
  - 2.4.4 Effect of Facet Reflectivity on Mode Selection — 26
- 2.5 Fundamental Device Characteristics — 28
  - 2.5.1 Light Intensity Versus Current — 28
  - 2.5.2 Optical Spectra — 29
  - 2.5.3 Near- and Far-Field Patterns — 30
- 2.6 Laser Diode to a Single-Mode Optical Fiber Coupling — 32
- 2.7 Direct Current Modulation of Semiconductor Lasers — 32
  - 2.7.1 Measurement of Small-Signal Modulation Response of a Laser Diode — 35
  - 2.7.2 Expression for the Relaxation Resonance Frequency — 37
- 2.8 Small-Signal Lumped-Element Equivalent Circuit Model — 38
- 2.9 Advanced Heterostructure Lasers — 38
  - 2.9.1 GaAs/GaAlAs Buried Heterostructure Laser — 38
  - 2.9.2 InP/InGaAsP Buried Heterostructure Laser — 40
  - 2.9.3 InP/InGaAsP Distributed Feedback Laser — 41
- 2.10 Reliability of Laser Diodes — 43
- References — 43

Chapter 3 ELECTRO-OPTIC MODULATORS — 47

- 3.1 Introduction — 47
- 3.2 Optical Waveguide Fundamentals — 48
  - 3.2.1 Single-Mode Optical Waveguides — 48
  - 3.2.2 Single-Mode Optical Branching Waveguide — 50
- 3.3 Electro-Optic Control — 51
  - 3.3.1 Index Ellipsoid — 51
  - 3.3.2 Electro-Optic Effect — 52
  - 3.3.3 Index Changes — 53
- 3.4 Electrode Structure — 54

3.5	Electrode Impedance and Attenuation		56
3.6	Coupling Loss Between a Single-Mode Optical Fiber and a Channel Waveguide		58
3.7	Interferometric Waveguide Modulators		59
	3.7.1	Principle of Operation	59
	3.7.2	Extinction Ratio	60
	3.7.3	Drive Voltage and Electrode Length Product	62
	3.7.4	Modulator Bandwidth and Electrode Length Product	63
	3.7.5	Techniques to Reduce Velocity Mismatch	64
	3.7.6	Optical Damage and dc Drift	65
	3.7.7	Radiation Hardness	66
	3.7.8	Experimental Modulator Performance and Discussions	66
	References		67
Chapter 4	PHOTODETECTORS		69
4.1	Introduction		69
4.2	PIN Photodiode Principle of Operation		70
	4.2.1	Quantum Efficiency and Photoresponsivity	71
	4.2.2	Response Speed	75
	4.2.3	Practical PIN Photodiode Structure	76
	4.2.4	Leakage Current	78
	4.2.5	Lumped-Element Equivalent Circuit Model and 3 dB Bandwidth	78
	4.2.6	Spectral Response	79
	4.2.7	Signal-to-Noise Ratio and Noise Equivalent Power	80
	4.2.8	Reliability	83
4.3	Planar Photoconductive Detector		83
	4.3.1	Photoconductive Gain and Carrier Lifetime	84
	4.3.2	Gain-Bandwidth Product and Rise time	87
	4.3.3	Signal-to-Noise Ratio and Noise Equivalent Power	88
	4.3.4	Practical Photoconductive Detector Structures	89
	4.3.5	Dark Current	90
	4.3.6	Lumped Element Equivalent Circuit Model	93
4.4	Schottky-Barrier Photodiodes		95
	4.4.1	Quantum Efficiency	96
	4.4.2	Response Speed and 3 dB Bandwidth	98
	4.4.3	Dark Current	98
	4.4.4	Practical Schottky-Barrier Photodiodes	100
	4.4.5	Lumped Element Equivalent Circuit Model	100
4.5	Avalanche Photodiodes		101

	4.5.1	Separate Absorption Grading and Multiplication-Avalanche Photodiode (SAGM-APD)	102
	4.5.2	Low Frequency Gain and Responsivity	102
	4.5.3	Frequency Response of the APD	103
	4.5.4	Signal-to-Noise Ratio and Noise Equivalent Power	105
4.6	Comparison of Photodetectors		106
4.7	Field-Effect Transistors		106
	4.7.1	Device Structure	107
	4.7.2	FET dc Characteristics Under Optical Illumination	111
	4.7.3	FET Microwave Characteristics Under Optical Illumination	114
	4.7.4	FET Noise Characteristics Under Optical Illumination	116
	References		117
Chapter 5	**MICROWAVE FIBER OPTIC LINKS**		**121**
5.1	Introduction		121
5.2	Insertion Loss of a Microwave Fiber Optic Link		123
	5.2.1	Directly Modulated Link	126
	5.2.2	Externally Modulated Link	126
5.3	Relative Intensity Noise of a Semiconductor Laser Diode		127
5.4	Noise Figure of a Microwave Fiber Optic Link		129
	5.4.1	Directly Modulated Link	129
	5.4.2	Externally Modulated Link	134
5.5	Link Bandwidth		139
5.6	Harmonic and Intermodulation Distortions		139
5.7	Spurious-Free Dynamic Range and Maximum Signal-to-Noise Ratio		142
5.8	Single-Mode Optical Fiber Dispersion		142
5.9	Optical Losses in a Microwave Fiber Optic Link		145
	5.9.1	Single-Mode Optical Fiber Attenuation	145
	5.9.2	Laser Diode to a Single-Mode Fiber Coupling Loss	146
	5.9.3	Connector Loss	146
	5.9.4	Splice Loss	147
	5.9.5	Coupler Loss	147
5.10	Performance Comparison of Microwave Fiber Optic Links		147
5.11	Fiber Optic Link Design Examples		147
	5.11.1	Directly Modulated, Medium Bandwidth, Long-Distance, Point-to-Point Microwave Link	147

		5.11.2	Directly Modulated, Narrow-Band, Short-Distance, Multiport Microwave Link	150
		5.11.3	Point-to-Point Microwave Link when Limited by Receiver Noise	151
		5.11.4	Externally Modulated, Short-Distance, Point-to-Point Microwave Link	151
		5.11.5	Computer Aided Design Modeling of a Laser Diode and a Photodiode for a Microwave Fiber Optic Link	153
	References			154
Chapter 6	OPTOELECTRONIC SWITCHING AND GATING			157
	6.1	Introduction		157
	6.2	Microstrip Optoelectronic Switch with Top-side Excitation		157
		6.2.1	Principle of Operation	157
		6.2.2	Gap Series Conductance	160
		6.2.3	Gap Shunt Conductance	160
		6.2.4	Insertion Loss in the On-State	161
		6.2.5	Repetition Rate	161
		6.2.6	Isolation in the Off-State	161
		6.2.7	Switching Speed	162
		6.2.8	Experimental Switch Performance and Discussions	162
	6.3	Optoelectronic Switching and Electronic Switching Performance Comparison		163
	References			164
Chapter 7	OPTOELECTRONIC MICROWAVE SIGNAL GENERATION			165
	7.1	Introduction		165
	7.2	Longitudinal Mode Power Spectra of a Free-Running Laser Diode		166
	7.3	Power Spectra of a Directly Frequency-Modulated Laser Diode		168
	7.4	FM Sideband Injection-Locking Technique		168
	7.5	Microwave Signal Generation Technique		169
	7.6	Experimental Generator Performance and Discussions		170
	7.7	Capabilities and Limitations		171
		7.7.1	Frequency Range	171
		7.7.2	Signal-to-Noise Ratio	171
		7.7.3	Temperature Stability	173
		7.7.4	Power Output and Efficiency	173
		7.7.5	Pulling Figure	174
		7.7.6	Stray Magnetic Field	174
		7.7.7	Shock and Vibrations	175
	References			175

Chapter 8 OPTOELECTRONIC SWITCH MATRIX 177
   8.1 Introduction 177
   8.2 Principle of an Optoelectronic Switch Matrix 178
   8.3 Avalanche Photodiode Optoelectronic Crosspoint Switch ... 178
       8.3.1 Principle of Operation 178
       8.3.2 Physical Origin of the Equivalent Circuit Model Elements 179
       8.3.3 Signal Transmission in the On-State 181
       8.3.4 Isolation in the Off-State 182
   8.4 Experimental Switch Performance and Discussions 182
   8.5 Optoelectronic Switch Matrix and Electronic Switch Matrix Performance Comparison 183
       8.5.1 Intermodulation Distortion 184
       8.5.2 Signal-to-Noise Ratio 185
       8.5.3 Radiation Hardness 189
       8.5.4 Reliability 192
      References 192

Chapter 9 OPTOELECTRONIC SWITCHING AND MODULATION OF OSCILLATORS 195
   9.1 Introduction 195
   9.2 Physical Mechanism of IMPATT Diode Operation 196
       9.2.1 Unilluminated IMPATT Diode 196
       9.2.2 Illuminated IMPATT Diode 198
   9.3 IMPATT Diode Oscillator Design 199
       9.3.1 Reduced Height Waveguide Cavity 199
       9.3.2 Standard Height Waveguide Cavity with Resonant Cap 201
   9.4 Experimental Demonstration of Enhancement and Quenching of IMPATT Diode Oscillator under Optical Illumination ... 202
       9.4.1 Frequency Chirp 203
   9.5 IMPATT Diode Structures with Etched Optical Window 203
   9.6 Effect of Hole *versus* Electron Photocurrent on Silicon IMPATT Diode Oscillator Power and Frequency 205
   9.7 Photocurrent Effects on Silicon IMPATT Oscillator Noise .. 205
   9.8 Frequency Modulation of a Silicon IMPATT Diode Oscillator by Optically Generated Carriers 206
   9.9 GaAs MESFET Oscillator Design 206
       9.9.1 Series Feedback Oscillator 206
   9.10 Experimental Demonstration of Switching of GaAs MESFET Oscillator under Optical Illumination 210

	9.11 Frequency Modulation of a GaAs MESFET Oscillator by Optically Generated Carriers	211
	References	211
Chapter 10	OPTOELECTRONIC INJECTION-LOCKING AND TUNING OF OSCILLATORS	213
	10.1 Introduction	213
	10.2 Brief Review of Conventional Electrical Injection-Locking Techniques and Experiments	214
	10.2.1 CW Oscillators	214
	10.2.2 Pulsed Oscillators	216
	10.3 Design of a Millimeter-Wave Microstrip IMPATT Diode Oscillator	217
	10.4 Design of a Microstrip GaAs MESFET Oscillator	220
	10.5 Direct Optical Injection-Locking and Tuning of CW Oscillators	223
	10.5.1 Principle of Operation	223
	10.5.2 Experimental Demonstration of Direct Optical Injection-Locking and Tuning	224
	10.6 Indirect Optical Injection-Locking of CW Oscillators	225
	10.6.1 Principle of Operation	225
	10.6.2 Experimental Demonstration of Indirect Optical Injection-Locking	226
	10.7 Comparison of Direct and Indirect Optically Injection-Locked CW Oscillator Performance	226
	References	228
Index		231

Foreword

Optics has existed as a standard element in the physics curriculum for over 100 years. Similarly, the younger microwave technology has been evolving for perhaps 50 years— as the exclusive province of the electrical engineer for most of that period. In the last 20 years, however, the two disciplines have continuously moved closer. In part this is the result of tremendous advances in solid-state physics. The most important of these are the solid-state lasers and detectors which made it possible to modulate optical signals at microwave frequencies. Simultaneously, the increased capabilities made feasible by such advances have been matched by the development of increasingly complex concepts in microwave systems. Advanced concepts in satellite communications systems, for example, feature on-board switching and data processing and overwhelmingly complex antennas and interconnect problems. The constraints of size, power, and freedom from interference imposed by any space system, simply reemphasize the need for a low-loss, lightweight interference-free transmission medium for which the microwave engineer is always searching.

Radar systems with phased array antennas and remote control stations similarly require complex interconnections. Electronic warfare and other weapons systems require bandwidth and security far beyond the capabilities of normal microwave transmission media. In some areas, particularly remote sensing applications, the traditional microwave or millimeter-wave experiments have been extended in frequency to where optical devices appear feasible as a means of generating the required signals for use as local oscillators. Simultaneously, the demands for data transmission capacity have reached the point where even the baseband digital data rates of several gigabits per second constitute a microwave problem.

It is clear that the disciplines of optics and microwaves are no longer distinct entities, and that there exist large areas (such as communications applications), where the knowledge of both technologies is at least enhancing and in many cases critical to the optimization of system performance. In most cases where two technologies merge, the researcher or system designer brings from his formal education an in-depth knowledge of one, but encounters an interdisciplinary gap which requires an introduction to the concepts and terminology of the other.

This book admirably bridges that gap. Whether the reader's primary expertise lies in the area of microwave electronics, solid state physics, or optics he will find that this book provides a good, practical exposure to the basic principles of the "deficient" area. The author has carefully organized the book according to the basic components most frequently encountered in microwave-optical systems. While the primary emphasis and most of the examples can be drawn from communications systems, the basic principles of the devices: signal generators, modulators, and switches, are universally applicable. In each case, the author has carefully described the basic physics of the topic and has given realistic numerical details as well as design examples. He has described devices, components and systems in terms which should make them understandable to us when approaching from either side of the gap. As a textbook, this book is useful at either the advanced undergraduate or graduate level. As a reference, it is valuable to either the researcher looking for an introduction to a new area or as a design guide for the worker in the field.

R.F. Leonard
NASA Lewis Research Center
August 30, 1989

Preface

This book is a comprehensive introduction to the theory and applications of optical control of microwave devices for graduate students in electrical engineering, and for practicing engineers who wish to improve their understanding of the principles and applications of this relatively recent development.

Optical control of microwave devices is a title used to describe one of several techniques for the control of microwave circuits. One approach is the simple replacement of coaxial cables and rectangular waveguides by optical fibers. Recently, however, another approach has become viable. Inasmuch as optical solidstate devices use the same materials (GaAs and InP), the same fabrication techniques, and the same processing techniques as microwave devices, it should be possible to integrate laser diodes, photodetectors, and modulators on a single chip with field-effect transistors to produce an *optical microwave monolithic integrated circuit* (OMMIC).

These two areas have traditionally been treated as separate disciplines, but the emergence of an OMMIC would create a need for an integrated approach to their study. This book begins with a chapter each on laser diodes, photodetectors, and electrooptic modulators. The remaining chapters are devoted to the detailed study of analog fiber optic links, optoelectronic switching and gating, optoelectronic microwave signal generation, an optoelectronic switch matrix, and optoelectronic switching, modulation, tuning and injection locking of microwave oscillators. The author includes many valuable details such as: the derivation of the fundamental equations, physical explanation, and numerical examples.

The book is an outgrowth of research work done by the author during a Research Associateship (1985–87) awarded by the National Research Council, Washington, DC, and carried out at the Solid State Technology Branch of the Lewis Research Center, National Aeronautics and Space Administration (NASA), Cleveland, Ohio. Since that time, interest in this topic among engineers has increased tremendously, with the concepts being extensively pursued by NASA for future space programs and missions. Thus, the decision to publish a book on the above subject appears appropriate.

In the course of writing this book, several persons have assisted the author and offered their support. The author first expresses his appreciation to the management of the Space Electronics Division at Lewis Research Center, for providing the environment in which he worked on the book; without their support, this book could not have materialized. The author has benefited significantly from the critical comments and constructive suggestions provided by Dr. R.F. Leonard throughout the preparation of the manuscript.

The author thanks Mr. Mark Walsh of Artech House, who suggested and encouraged the writing of this book, the editorial staff for the processing of the manuscript, and the anonymous reviewers whose suggestions improved the content of the book.

At the Solid State Technology Branch, the author is grateful to the engineers and scientists who shared their time, knowledge and understanding of this subject. In addition, his sincere thanks go to Ms. K.J. McKee for her timely secretarial assistance in the preparation of the manuscript.

At the Lewis Research Center, the author is grateful to the members of the Graphics Section and the Photolaboratory for their efficiency in the preparation of the illustrations.

Finally, the author thanks his wife, Joy, and daughters, Renita and Rona, for their patience during the writing of this book.

R.N. Simons
Cleveland, Ohio
September 5, 1989

Chapter 1
Introduction

In the past few years tremendous strides have been made in the realization of *microwave monolithic integrated circuits* (MMICs) for communications, radar and high-speed digital applications [1]. This has been made possible for several reasons: (1) the availability of III-V compound semiconductors such as gallium arsenide (GaAs) and indium phosphide (InP), which have very high electron mobility and scatter limited velocity, (2) the ability to grow semi-insulating substrates on which it is possible to fabricate low-loss planar transmission lines such as microstrip for interconnection, (3) the availability of several techniques such as *molecular-beam epitaxy* (MBE) and *vapor or liquid phase epitaxy* (VPE, LPE) for the epitaxial growth of ternary and quaternary compounds such as GaAlAs and InGaAsP to form multilayer heterostructures, (4) the availability of test equipment for on-wafer characterization of devices and (5) computer aided design software for device performance optimization. As a result of these advances, discrete field-effect transistors with unity current-gain cutoff frequency (f_t) greater than 125 GHz [2–4] and noise figure and gain of about 1.5 dB and 6 dB respectively at 94 GHz have been achieved. In addition, advances have also been made in realizing special IMPATT (impact avalanche and transit time) diode [5] and PIN (positive-intrinsic-negative) diode [6] structures suitable to monolithic integration as well as for efficient coupling of light into the device active region.

Concurrently, there has been considerable progress made in the development of optoelectronic devices such as laser diodes, photodetectors and modulators. Laser diodes with relaxation frequencies as high as 30 GHz [7] have been reported recently, and there are indications that this limit will become much higher [8]. Photodetectors with bandwidth greater than 100 GHz [9] have been demonstrated. Electro-optic intensity modulators of the Mach-Zehnder type [10] and the traveling-wave type [11] with bandwidths on the order of 6 GHz and 20 GHz, respectively, have been realized. Advances in fabrication and processing techniques should eventually lead to further improvements in design and reduction of parasitic capacitances, thereby extending the upper limit on bandwidth into the millimeter-wave region. Similar advances in the development of *optical* monolithic integrated circuits (OMICs) [12, 13] for optical communications have also occurred. The semiconductor materials from which these emitters, detectors and modulators have been realized are GaAs, InP, $In_xGa_{1-x}As$, $Ga_{1-x}Al_xAs$ and $Ga_{1-x}In_xAs_yP_{1-y}$, where x and y are the adjustable mole fraction. These advances have taken place typically in the wavelength range of 0.8 to 0.85 μm and 1.3 to 1.55 μm. The longer wavelength is of greater interest because optical fibers have the least loss at these wavelengths, typically less than 0.5 dB/km, and almost zero dispersion, which results in a bandwidth of several hundred GHz-km.

Traditionally, monolithic microwave integrated circuits and optical-monolithic integrated circuits were considered to be two separate fields of study. However, the fact that the same material system is used to produce both microwave and optical devices and integrated circuits suggests that these two areas of research can be combined to produce optical *microwave* monolithic integrated circuits (OMMICs).

In this book, we present various techniques for the optical control of microwave devices. Several of these techniques have not yet been demonstrated in the monolithically integrated form, but they have been demonstrated in discrete or hybrid configurations. Nevertheless, the same basic principles apply, and monolithic integration will inevitably follow.

1.1 ADVANTAGES OF OPTICAL MICROWAVE MONOLITHIC INTEGRATED CIRCUITS

1.1.1 Performance

Monolithic integration reduces the parasitic reactances associated with the packages and bonding wires and in general improves the circuit response. For example, in the case of photoreceivers, a reduction in the input-parasitic capacitance improves both the speed and the signal-to-noise characteristics resulting in higher receiver sensitivity.

In applications involving direct high-speed current modulation of laser diodes, one of the problems encountered is frequency chirping. The frequency chirping together with the dispersion of the optical fiber results in the deterioration of the link transmission characteristics. By integrating an external modulator with a laser diode frequency, chirping can be minimized.

Conventional Fabry-Perot laser diodes have a broad spectral line-width which renders them useless for coherent communications. One of the techniques to obtain a narrow line-width source is to use a distributed-feedback laser which makes use of a Bragg reflection grating. A simpler technique would be to integrate an external cavity with the laser.

The last performance advantage obtained from integration is the reduction in the delay time, due to interconnection between circuits.

1.1.2 Design

Because optical sources, detectors and transistors all have planar topology they can all be laid out and interconnected by planar transmission lines. This feature permits integration of devices while retaining the capability to optimize individual devices. Integration also results in uniformity in the circuit characteristics. Finally, due to the inherent isolation between optical and microwave signals special care need not be exercised to prevent crosstalk.

1.1.3 Manufacturing

The advantages gained in the manufacturing process by optical-microwave integration are those associated with most integrated circuits: simple assembly, high productivity, compactness, reliability and low cost.

1.2 TYPICAL OPTICAL MICROWAVE MONOLITHIC INTEGRATED CIRCUITS

1.2.1 Laser Diode and Field-Effect Transistor

A crucial step in the design of integrated circuits is the fabrication of low-threshold-current laser diodes. A lower threshold current results in lower heat dissipation and hence increased reliability. Furthermore, a lower threshold enables the laser diode to be driven to higher current levels thereby extending the relaxation oscillation frequency and consequently, the frequency bandwidth. One such low threshold (8 to 15 mA) GaAlAs/GaAs buried heterostructure laser has been monolithically integrated with a *metal-semiconductor field-effect transistor* (MESFET). The circuit has a direct modulation bandwidth of about 4 GHz under CW operation [14].

1.2.2 Laser Diode and Gunn Diode

Gunn diodes are an important source of microwave power. Hence, monolithic integration of a GaAlAs/GaAs injection laser and a Gunn diode has been undertaken [15]. In a practical circuit, the two devices are arranged in series so that the oscillating current pulses from the Gunn diode are injected to the laser diode, intensity modulating its light output. Using this circuit, intensity modulation of light at a frequency as high as 1 GHz has been achieved.

1.2.3 Laser Diode and Heterojunction Bipolar Transistor

The InGaAsP/InP laser is of tremendous interest for long wavelength, high-bit-rate optical communications. Also, *heterojunction bipolar transistors* (HBT) are emerging as an alternative to MESFETs because they are potentially faster, less noisy at high frequency and have better device-to-device uniformity [16]. These two devices have been monolithically integrated and current-controlled optical switching has been demonstrated [16].

1.2.4 Laser Diode and an Optical Intensity Modulator

The advantages of integrating a laser diode and an external modulator were pointed out earlier in Section 1.1.1. In the best approach demonstrated to date, the modulator uses a quantum well formed by layering an ultrathin, lower band-gap semiconductor between two lattice-matched, wide band-gap semiconductors. The first successful monolithic integration of such a *multiquantum well* (MQW) optical modulator and an InGaAsP/InP *distributed feedback laser* (DFB) has been reported [17]. The new circuit is capable of achieving as high as 55% modulation depth.

Recent studies indicate that a bandwidth as high as 8 GHz [18] can be achieved with the above scheme. In addition, the chirp reduces to less than 0.1 nm, which is almost an order of magnitude smaller than that for a directly modulated semiconductor laser [18].

1.2.5 Metal-Semiconductor-Metal Photodetector and Field-Effect Transistor

The advantages of integrating a photodetector and the post detection amplifier were also pointed out earlier in Section 1.1.1. The first successful monolithic receiver that uses sub-half-micrometer GaAs MESFETs and metal-semiconductor-metal photodetector has been reported [19]. The receiver has a bandwidth of 5.2 GHz and a transimpedance bandwidth product as high as 1.5 THz Ω.

1.2.6 PIN Photodetector and Field-Effect Transistor

Monolithic integrated photoreceivers consisting of an InGaAs PIN photodiode and a transimpedance amplifier were reported in [20]. The key feature of this receiver is its ability to operate from a 5 V power supply. This results in considerable simplification of the system insofar as most digital integrated circuits operate with a 5 V power supply. In addition, easily producible device structures were adopted in order to increase the yield of the photoreceiver, which is essential to make optoelectronic circuits practical.

1.2.7 Laser Diode, Photodetector and Field-Effect Transistor

The first successful fabrication of a AlGaAs/GaAs multichannel (four) monolithic integrated optoelectronic transmitter consisting of a laser diode, photodetector, and a *field-effect transistor* (FET) was reported in [21]. The quantum efficiency of the laser diode is in the range of 50 to 60%, the responsivity of the photodetector is in the range of 1.8 to 3 μA/mW, and the transconductance of the FET is typically about 50 mS/mm. This resulted in an overall input-voltage to output-light power conversion ratio of about 6 mW/V, and a modulation capability of greater than 1 Gb/s. The overall characteristics have been shown to be uniform over the four channels.

1.3 TYPICAL APPLICATIONS

1.3.1 Beam Steering in Phased Arrays

Future generations of phased array radar systems as well as satellite-borne communication systems might need several thousand active radiating elements to form a pencil beam for tracking and communication, respectively. In addition, advanced phased array radars for tactical aircraft will require distributed transmitter-receiver modules constructed from GaAs MMICs and arranged in an antenna architecture that conforms to the skin of the fuselage [22]. The beam steering in these arrays is done electronically by rapidly varying the relative phase of the radiating elements. This can be accomplished optically by using either a fiber optic link to carry digital commands to the MMIC phase shifters [23], or optical fibers to perform RF power distribution and phase shifting [24]. At NASA Lewis Research Center, a monolithic optical interface chip has been developed under contract, for a phased array system which uses a fiber optic link to carry digital commands to MMIC modules [23]. Figure 1.1 shows the schematic of the monolithic-optical MMIC control interface chip. In this circuit the optical receiver converts the optical serial data stream into a 1 Gb/s electrical data stream. The 1:16 demultiplexer converts the logic-level serial data stream into its time demultiplexed 16-bit data components.

1.3.2 Reference Frequency Distribution in Phased Arrays

The individual MMIC transmit-receive modules in a large phased array antenna are independent of one another and must be synchronized to a master oscillator for the radiation to combine coherently and form a single beam in free space. Synchronization may be achieved by injection locking the slave oscillators in the modules to the master oscillator. Conventional techniques dictate the use of coaxial cables, however, the preferred way would be to use a fiber optic link to carry the reference signal because of its size and weight advantages.

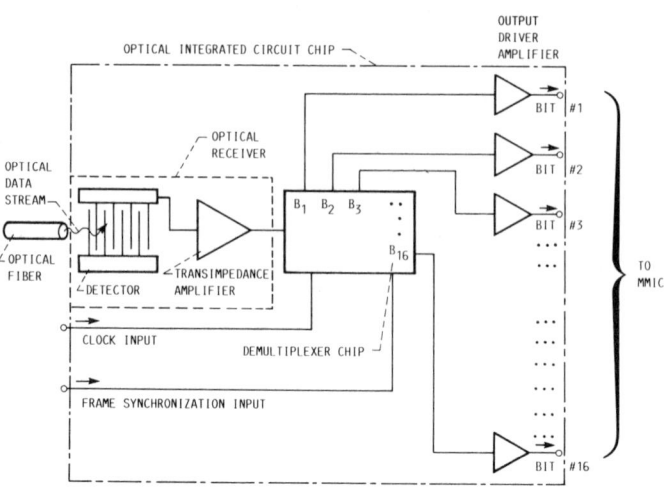

Figure 1.1 Schematic of the optical monolithic integrated circuit control interface chip.

1.3.3 Remoting of Antennas

The antenna systems of modern radars for weapons control present a high profile visible target and also act as an electromagnetic emission source. Thus, the antenna system is vulnerable to enemy missile attack. Therefore, operating personnel and the expensive control equipment should be remotely located from the antenna. The conventional technique of linking the antenna to the command center is to use coaxial cables. Insofar as the conductor loss in any transmission line increases as the square root of frequency, to down-convert the microwave signals to the MHz range is usually necessary before transmission through coaxial cables. In addition, several line amplifiers are needed to amplify the signal level. However, if fiber optic links are used then transmission of the microwave signal to the command center can be done without any down-conversion. Furthermore, there is no need for any line amplifiers to boost the signal level because the attenuation of optical fibers is extremely small.

1.3.4 Signal Processing and Electronic Warfare

Signal processing in radar systems and electronic warfare frequently require broadband delay lines [25]. The conventional method of achieving these delays is to use coaxial cables or bulk acoustic-wave devices. The transmission delay of coaxial cables is about 5 ns/m. Considering the size and weight of a typical coaxial cable a feasible delay is on the order of 200 to 300 ns. Bulk acoustic-wave devices on lithium niobate extend this range to about 5 µs [26]. A silica fiber provides about 5 µs of delay per kilometer of length. Because several kilometers of fibers can be coiled into a small volume, delays of up to 300 µs are feasible [26].

1.3.5 Missile Guidance

A novel application of fiber optic links is in guided missiles [27], [28]. The fiber optic link is used for transmitting in-flight imagery to the launch unit located as far as 10 km [34]. Because fiber optic links are impervious to EMI, the missile would be extremely difficult to jam.

1.3.6 Cable Television

In cable television, the earth station antenna for reception from satellites must be located so that it has minimal terrestrial interference, has a full view of the arc of geosynchronous satellites, and is aesthetically acceptable. Therefore, an appropriate location for the earth station antenna would be on the outskirts of a city. However, the head-end must be located where the cable television output can be easily distributed to the subscribers. One way to connect the earth station antenna to the head-end is to use an analog fiber optic link [29].

1.3.7 Cellular Telephone

A recent application of fiber optics has been to extend cellular telephone coverage to shadow zones, or sites that are obscured from the main antenna for that local cell. Such sites include bridges, underground parking structures, tunnels and the far sides of tall buildings. Fiber optic links can be run from the nearest transmission node to an antenna within the shadow area [29].

1.3.8 Instrumentation

A fiber optic link has been used in an antenna near-field test range at NASA Lewis Research Center [30]. The link delivers the local oscillator (LO) signal (2 to 6 GHz) to the harmonic mixers in the receiving antenna. A particular requirement of near-field measurement is phase stability. Two limitations on phase stability imposed by using coaxial cables have been overcome by the use of optical fibers. First, the instability associated with flexure of the cable during measurement is eliminated. Second, the requirement for a rotary cable joint is avoided. Thus, the dynamic range, sensitivity and phase measurement accuracy are vastly improved. Fiber optic links have also found application in conventional anechoic-chamber antenna far-field measurements [31].

Picosecond photoconductivity sampling techniques have also proven to be useful in the measurement of the far-field radiation pattern of monolithically-integrated antennas [32]. Traditionally, large and expensive anechoic chambers were used to suppress unwanted reflections from surroundings, while the pattern was measured. In the photoconductive technique, the waveform of interest is time-gated while unwanted reflections arriving late at the receiver are rejected. Consequently, discrimination of reflections from structures within a few millimeters of the antenna under test can be accomplished by the use of sufficiently short pulses. In addition, investigation of device performance at frequencies approaching 1 THz is feasible.

Another interesting application of fiber optic links is in the phase noise measurement of radars [33]. This is made possible by the large delay times that can be realized using fibers.

1.4 ORGANIZATION OF THIS BOOK

This book is organized to serve as a text for a graduate course in optical control of microwave devices and circuits, as well as a reference volume for scientists and engineers in industry. Hence, Chapter 1 gives an overview of the advantages and typical examples of OMMICs.

Chapters 2 through 4 are devoted to the basic devices such as laser diodes, photodetectors and intensity modulators which tie the OMMICs together. Therefore, the basic principles and characteristics of heterostructure laser diodes, as well as several variants of the basic structure are explained in detail in Chapter 2. Chapter 3 discusses the principles of a Mach-Zehnder electro-optic intensity modulator fabricated on $LiNbO_3$. Chapter 4 presents the device structures, principle of operation, and characteristics of PIN, photoconductive, Schottky-barrier, and avalanche photodiodes. In addition, we also discuss the microwave performance of optically illuminated MESFETs and HEMTs.

Chapter 5 discusses the principles of microwave fiber optic links. The advantages and disadvantages of directly modulated links as well as externally modulated links are compared. The Chapter concludes with several examples of practical fiber optic links.

Optoelectronic switches have several advantages in high-power switching and also in instrumentation applications. Hence, Chapter 6 is devoted to optoelectronic switches and gates constructed from planar transmission lines such as microstrip and *coplanar waveguides* (CPW). The characteristics of this switch are explained and its performance compared with that of conventional electronic switches.

Due to size and weight limitations of coaxial cables and rectangular waveguides, there is a need to explore the use of optical fiber and semiconductor laser diode technologies for the generation and distribution of microwave signals in future active phased arrays. Therefore, Chapter 7 discusses optoelectronic techniques to generate a microwave signal. The capabilities and limitations of this technique are also presented.

A switch matrix is an important subsystem in some multibeam satellite architectures, providing rapid interconnectivity between earth stations. An optoelectronic switch matrix based on avalanche photodiodes or PIN photodiodes is discussed in Chapter 8. This chapter also compares the performance of an optoelectronic switch matrix and an electronic switch matrix.

Chapter 9 is devoted to the optoelectronic switching and modulation of IMPATT and MESFET oscillators. In addition, the chapter presents the design of IMPATT as well as MESFET oscillators. Finally, frequency modulation of these oscillators is discussed.

Finally, Chapter 10 presents optoelectronic injection locking and tuning of microwave oscillators. The advantages and disadvantages of both direct optical injection-locking as well as indirect optical injection-locking are presented. In both Chapters 9 and 10, wherever possible, a performance comparison is made with conventional electronic control of oscillators.

REFERENCES

1. Pucel, R.A. (ed.), Monolithic Microwave Integrated Circuits, IEEE Press, New York, 1985.
2. Chen, Y.K., R.N. Nottenburg, M.B. Panish, R.A. Hamm, and D.A. Humphrey, "Subpicosecond InP/InGaAs Heterostructure Bipolar Transistors," *IEEE Electron Device Letters*, Vol. 10, No. 6, June 1989, pp. 267–269.
3. Wang, G.W., and M. Feng, "Quarter-Micrometer Gate Ion-Implanted GaAs MESFETs with an f_t of 126 GHz," *IEEE Electron Device Letters*, Vol. 10, No. 8, August 1989, pp. 386–388.
4. Nguyen, L.D., D.C. Radulescu, P.J. Tasker, W.J. Schaff, and L.F. Eastman, "0.2 μm Gate-Length Atomic-Planar Doped Pseudomorphic $Al_{0.3}Ga_{0.7}As/In_{0.25}Ga_{0.75}As$ MODFETs with F_T over 120 GHz," *IEEE Electron Device Letters*, Vol. 9, No. 8, August 1988, pp. 374–376.
5. Stabile, P.J., and B. Lalevie, "Lateral IMPATT Diodes," *IEEE Electron Device Letters*, Vol. 10, No. 6, June 1989, pp. 249–251.
6. Stabile, P.J., A. Rosen, and P.R. Herczfeld, "Optically Controlled Lateral PIN Diodes and Microwave Control Circuits," *RCA Review*, Vol. 47, December 1986, pp. 443–456.
7. Uomi, K., T. Mishima, and N. Chinone, "Ultrahigh Relaxation Oscillation Frequency (up to 30 GHz) of Highly p-doped GaAs/GaAlAs Multiple Quantum Well Lasers," *Applied Physics Letters*, Vol. 51, No. 2, July 1987, pp. 78–80.
8. Lau, K.Y.,"Efficient Narrow-Band Direct Modulation of Semiconductor Injection Lasers at Millimeter-Wave Frequencies of 100 GHz and Beyond," *Applied Physics Letters*, Vol. 52, No. 26, June 1988, pp. 2214–2216.
9. Zeghbroeck, B.J.V., W. Patrick, J.M. Halbout, and P. Vettiger, "105-GHz Bandwidth Metal-Semiconductor-Metal Photodiode," *IEEE Electron Device Letters*, Vol. 9, No. 10, October 1988, pp. 527–529.
10. Walker, R.G., "Broadband (6 GHz) GaAs/AlGaAs Electro-Optic Modulator with Low Drive Power," *Applied Physics Letters.*, Vol. 54, No. 17, April 1989, pp. 1613–1615.

11. Wang, S.Y., S.H. Lin, and Y.M. Houng, "GaAs Traveling-Wave Polarization Electro-optic Waveguide Modulator with Bandwidth in Excess of 20 GHz at 1.3 μm," *Applied Physics Letters*, Vol. 51, No. 2, July 1987, pp. 83–85.
12. Chaim, N.B., I. Ury, and A. Yariv, "Integrated Optoelectronics," *IEEE Spectrum*, Vol. 19, No. 5, May 1982, pp. 38–45.
13. Forrest, S.R., "Optical Detectors: Three Contenders," *IEEE Spectrum*, Vol. 23, No. 5, May 1986, pp. 76–84.
14. Ury, I., K.Y. Lau, N.B. Chaim, and A. Yariv, "Very High Frequency GaAlAs Laser-Field Effect Transistor Monolithic-Integrated Circuit," *Applied Physics Letters*, Vol. 41, No. 2, July 1982, pp. 126–128.
15. Lee, C.P., S. Margalit, I. Ury, and A. Yariv, "Integration of an Injection Laser with a Gunn Oscillator on a Semi-Insulating GaAs Substrate," *Applied Physics Letters*, Vol. 32, No. 12, June 1978, pp. 806–807.
16. Chen, T.R., K. Utaka, Y. Zhuang, Y.Y. Liu, and A. Yariv, "A Vertical Monolithic Combination of an InGaAsP/InP Laser and a Heterojunction Bipolar Transistor," *IEEE J. Quantum Electronics*, Vol. QE-23, No. 6, June 1987, pp. 919–924.
17. Kawamura, Y., K. Wakita, Y. Yoshikuni, Y. Itaya, and H. Asahi, "Monolithic Integration of a DFB Laser and an MQW Optical Modulator in the 1.5 μm Wavelength Range," *IEEE J. Quantum Electronics*, Vol. QE-23, No. 6, June 1987, pp. 915–918.
18. Soda, H., M. Furutsu, K. Sato, M. Matsuda, and H. Ishikawa, "Dynamic Characteristics of an Optical Intensity Modulator Monolithically Integrated with a DFB Laser under 5-Gbits Modulation," *Conf. Digest Lasers and Electro-Optics (CLEO)*, April 1989, pp. 376–377.
19. Harder, C.S., B.V. Zeghbroeck, H. Meier, W. Patrick, and P. Vettiger, "5.2 GHz Bandwidth-Monolithic GaAs Optoelectronic Receiver, *IEEE Electron Device Letters*, Vol. 9, No. 4, April 1988, pp. 171–173.
20. Matsuda, K., M. Kubo, K. Ohnaka, and J. Shibata, "A Monolithically Integrated InGaAs/InP Photoreceiver Operating with a Single 5 V Power Supply," *IEEE Trans. Electron Devices*, Vol. 35, No. 8, August 1988, pp. 1284–1287.
21. Kuno, M.T. Sanada, H. Nobuhara, M. Makiuchi, T. Fujii, O. Wada, and T. Sakurai, "Four-Channel AlGaAs/GaAs Optoelectronic Integrated Transmitter Array," *Applied Physics Letters*, Vol. 49, No. 23, December 1986, pp. 1575–1577.
22. Scott, W.B., "Air Force Funding Joint Studies to Develop Smart Skin Avionics," *Aviation Week and Space Technology*, Vol. 128, No. 16, April 18, 1988, pp. 65–69.
23. Gustafson, G., M. Bendett, J. Carney, R. Mactaggart, S. Palmquist, F. Schmit, K. Tan, and W. Walters, "GaAs Circuits for Monolithic Optical Controller," SPIE, Vol. 886, *Optoelectronic Signal Processing for Phased-Array Antennas*, 1988, pp. 80–87.
24. "USAF Studies Linking Phased-Array Radar with Fiber Optic Cable," *Aviation Week and Space Technology*, Vol. 130, No. 5, January 30, 1989, pp. 61–63.
25. Pan, J.J., "Fiber Optics and Opto-Electronics for Radar and Electronic-Warfare Applications," *Microwave Systems News and Communication Technology*, Vol. 17, No. 10, October 1987, pp. 30–43.
26. Taylor, H.F., "Applications of Guided-Wave Optics in Signal Processing and Sensing," *Proc. IEEE*, Vol. 75, No. 11, November 1987, pp. 1524–1535.
27. "U.S., European Firms Seek to Develop Fiber Optic-Guided Antitank Missiles," *Aviation Week and Space Technology*, Vol. 128, No. 25, June 20, 1988, p. 111.
28. "French, Germans Test Fiber Optic Missile," *Aviation Week and Space Technology*, Vol. 128, No. 12, March 21, 1988, p. 32.
29. Grimes, G., "Where Microwave and Optics Meet," *Photonics Spectra*, Vol. 23, No. 5, May 1989, pp. 101–110.

30. "Microwave Fiber Optics-NASA Antenna Test Range Uses Lasertron Q-Links," Lasertron, Inc.
31. Kemp, W.M., and A.T. Tickner, "A Role of Fiber Optics in Antenna Measurements," *J. Electrical and Electronics Engineering* (Australia), Vol. 7, No. 4, December 1987, pp. 278–281.
32. Lutz, C.R., and A.P. DeFonzo, "Far-Field Characteristics of Optically-Pulsed Millimeter Wave Antennas," *Applied Physics Letters,* Vol. 54, No. 22, May 1989, pp. 2186–2188.
33. Newberg, I.L., C.M. Gee, G.D. Thurmond, and Y.H. Yen, "Radar Applications of X-Band Fiber Optic Links," *IEEE Int. Microwave Symp. Digest,* June 1988, pp. 987–990.
34. "U.S. Navy Tests Fiber-Optic Data Links for Air-Launched Weapons," *Aviation Week and Space Technology,* Vol. 130, No. 24, June 12, 1989, pp. 275–278.

Chapter 2
Laser Diodes

2.1 INTRODUCTION

In the rapidly developing field of optical communication, the semiconductor laser diode has become the most promising source of optical signal. The main advantages of the semiconductor laser diode over the gas laser is that it consumes only a small amount of dc power (current) for pumping, it has a higher efficiency (typically 10%), and is capable of being directly modulated up to the GHz range [1]. The early versions of the laser diodes were constructed from ternary semiconductors such as GaAlAs lattice matched to GaAs. These lasers emitted in the wavelength range of 0.8 to 0.9 μm [2]. The recent versions are fabricated from quaternary semiconductors such as InGaAsP lattice matched to InP. These lasers emit in the wavelength range of 1.1 to 1.6 μm [2]. This wavelength is of tremendous interest for optical communication insofar as the attenuation of single-mode optical fibers at 1.3 μm and 1.55 μm wavelength is as low as 0.5 dB/km and 0.2 dB/km, respectively [3], [4]. Furthermore, single-mode fibers have a dispersion minimum at about 1.29 μm wavelength resulting in a pulse broadening of only 4 ps/(nm.km) [5].

This chapter is organized as follows: Section 2.2 explains the basic principles of a semiconductor laser diode. There are several early versions of semiconductor laser diodes, however, the double heterostructure laser has proved to be the most promising device. Hence Section 2.3 presents the double heterostructure laser diode. Section 2.4 discusses the waveguiding principles and the intensity pattern of this device. Furthermore, Section 2.5 explains the fundamental device characteristics. Section 2.6 discusses laser diode to single-mode optical fiber coupling. Section 2.7 explains direct intensity modulation of a laser diode by a microwave signal. A small-signal, lumped-element equivalent circuit model of the laser diode which is useful for circuit design is discussed in Section 2.8. Lastly, Section 2.9 presents several advanced laser diodes which are capable of being directly modulated at microwave frequency.

2.2 THE CONCEPT OF STIMULATED EMISSION

Three vital processes give rise to laser action. These processes are: photon absorption, spontaneous emission, and stimulated emission. These three processes are illustrated in Figure 2.1 by a simple two-energy-level diagram. Planck's law states that a transition between energy levels involves either the absorption or emission of a photon. Mathematically, this can be represented as

$$h\nu = E_2 - E_1 \qquad (2.1)$$

where

E_1 = ground-state energy;
E_2 = excited-state energy;
$h\nu_{12}$ = photon energy.

In Figure 2.1(a), an electron from the equilibrium energy state is excited to a higher energy state by the absorption of a photon of energy $h\nu_{12}$. Since this is an unstable state, the electron will shortly return to the equilibrium state, thereby emitting a photon of energy $h\nu_{12}$ (Fig. 2.1(b)). This occurs without any external stimulation

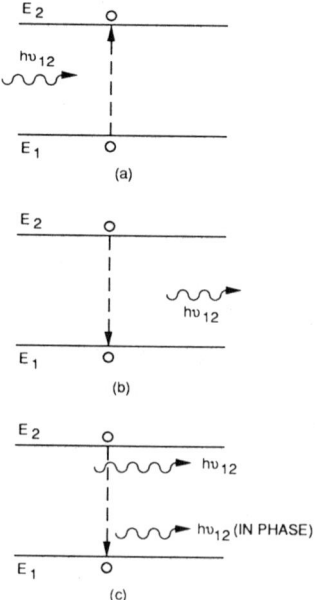

Figure 2.1 The three vital transition processes in laser action: (a) absorption; (b) spontaneous emission; (c) stimulated emission.

and is called *spontaneous emission*. These emissions are isotropic and of random phase. The electrons can also be induced to make a downward transition from the higher energy state to the equilibrium state by an external stimulation. As shown in Figure 2.1(c), if a photon of energy $h\nu_{12}$ is incident while the electron is still in the excited state, the electron drops to the equilibrium state and emits a photon of energy $h\nu_{12}$. These emissions are in phase with the incident photon and the resulting emission is known as *stimulated emission*.

For systems that are in thermal equilibrium, the density of excited electrons is very small. Therefore, such systems will absorb most of the incident photons and result in negligible stimulated emission. Stimulated emission will exceed absorption only if the population of the excited states is greater than that of the equilibrium state. This condition is known as *population inversion*. In semiconductor lasers population inversion is accomplished by injecting electrons into the material at the device contacts, to fill the lower energy states of the conduction band.

In order for electrons to transit to and from the conduction band with the absorption or emission of photons, respectively, both energy and momentum must be conserved. A photon can have considerable energy, but its momentum $h\nu/c$ is very small. Semiconductors such as GaAs and InP with which we are mainly concerned in this chapter are classified as direct band-gap materials. The shape of the band gap as a function of the momentum, k is illustrated in Figure 2.2. In this material, recombination of an electron and a hole having the same momentum value takes place.

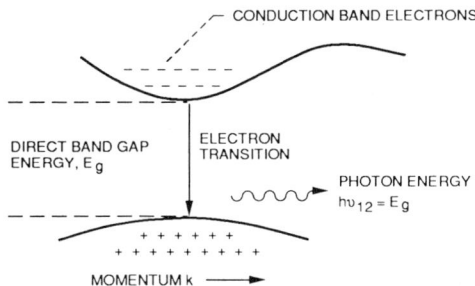

Figure 2.2 Electron recombination and the associated photon emission for a direct-band-gap semiconductor.

2.3 THE DOUBLE HETEROSTRUCTURE LASER DIODES

A junction between two materials of different band gaps such as GaAs and GaAlAs is known as a heterojunction and the resulting structure is known as a heterostructure. This basic structure is grown epitaxially on a crystalline GaAs substrate so that it is crystallographically uninterrupted. If two such junctions are required in a device, then it is called a *double heterostructure*. A typical double heterostructure GaAs/GaAlAs laser is shown in Figure 2.3. In this device a very thin GaAs active layer is sandwiched between a p-type $Ga_{1-y}Al_yAs$ and an n-type $Ga_{1-x}Al_xAs$ cladding layers. The difference between the indices of refraction of GaAs and the ternary $Ga_{1-x}Al_xAs$ with a molar fraction x and $Ga_{1-y}Al_yAs$ with a molar fraction y gives rise to a three-layered dielectric waveguide. The fundamental mode in this dielectric waveguide has its energy concentrated mostly in the high index GaAs layer. The index distribution and a typical modal intensity plot for the fundamental mode is shown in Figure 2.4.

2.3.1 Energy Band Structure

When a positive bias is applied to the device, electrons are injected from the n-type GaAlAs and holes are injected from the p-type GaAlAs into the active GaAs region. The electrons which are injected into the active region are prevented from diffusing out into the p-type GaAlAs by means of the potential barrier due to the difference ΔE_g between the energy gaps of GaAs and $Ga_{1-y}Al_yAs$. Similarly the difference between the energy gaps of GaAs and $Ga_{1-x}Al_xAs$ prevents holes from diffusing into the n-type GaAlAs. The x or the y dependence of the energy gap is approximated by [6]

$$E_g \,(x \text{ or } y < 0.37) = (1.424 + 1.247\,x), \text{ eV} \qquad (2.2a)$$

The peak emission wavelength λ in micrometers can be expressed as a function of the band gap energy E_g in electron volts by the equation:

$$\lambda \,(\mu m) = \frac{1.240}{E_g(\text{eV})} \qquad (2.2b)$$

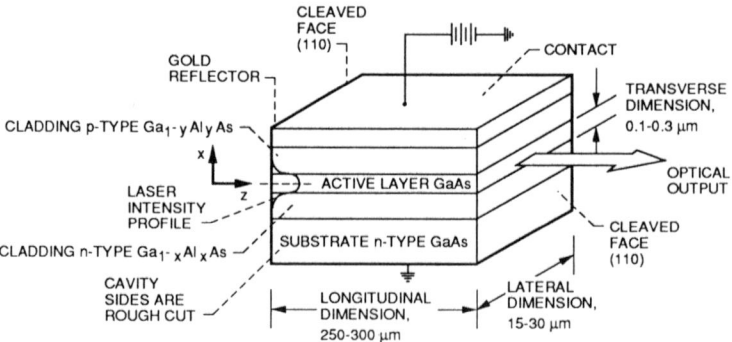

Figure 2.3 A typical GaAs-GaAlAs double heterostructure laser diode.

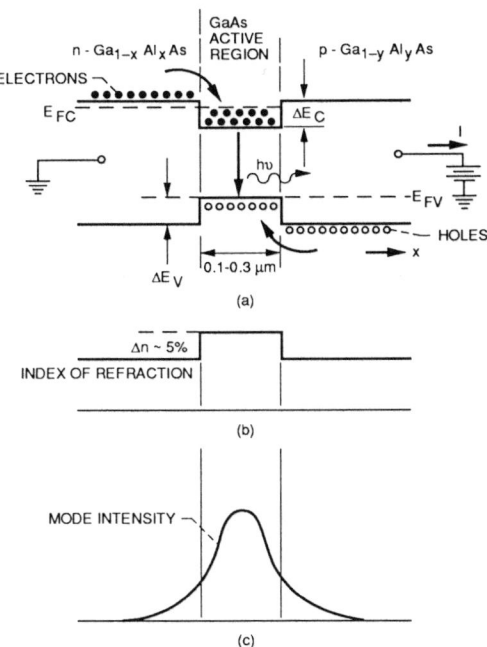

Figure 2.4 Forward-biased double heterostructure GaAs-GaAlAs laser diode: (a) energy band; (b) index of refraction; (c) intensity profile.

For example, $Ga_{0.93}Al_{0.07}As$ has $E_g = 1.51$ eV and emits at $\lambda = 0.82$ μm. The discontinuity at the conduction band and the valence band edges amount to mostly 85 and 15% of the total discontinuity ΔE_g at a GaAs/GaAlAs interface. This discontinuity effectively confines both holes and electrons to the active region. This double confinement of injected carriers, as well as the optical mode energy to the same region, results in a low-threshold continuously operated room temperature semiconductor laser.

Similarly, for $In_{1-x}Ga_xAs_yP_{1-y}$ lattice matched to InP the band gap is given by

$$E_g = 1.35 - 0.72\, y + 0.12\, y^2 \qquad (2.2c)$$

The compositional parameters x and y follow the relationship $y \approx 2.20x$ with $0 \leq x \leq 0.47$. The commonly used alloy composition in 1.3 μm lasers ($E_g = 0.96$ eV) is $In_{0.74}Ga_{0.26}As_{0.6}P_{0.4}$, whereas for 1.55 μm devices, it is $In_{0.6}Ga_{0.4}As_{0.9}P_{0.1}$.

2.3.2 Threshold Current Density

The internal quantum efficiency η in the active region of a laser diode is defined as the fraction of electron-hole pairs that recombine radiatively and is expressed as

$$\eta = \frac{\tau_c}{\tau} \tag{2.3a}$$

where

τ_c = carrier lifetime;
τ = spontaneous (radiative) recombination lifetime.

In a semiconductor laser diode, the gain g (the incremental energy flux per unit length) depends on the energy-band structure. The gain can also be determined as a function of the nominal current density J_{nom}, which is defined for unity quantum efficiency as the current density required for uniformly exciting a 1 μm thick active layer [7]. The actual current density is then given by [8]

$$J = \frac{J_{nom} d}{\eta} \quad \text{A/cm}^2 \tag{2.3b}$$

where

d = active layer thickness
η = internal quantum efficiency.

The variation of the gain, g for a uniformly excited semiconductor with the nominal current density, J_{nom} has been analytically determined for GaAs at a fixed temperature and ionization impurity concentration [7] and is expressed as

$$g = \frac{g_0}{J_0}(J_{nom} - J_0) \tag{2.4}$$

In the above equation, typical values for the constants are [8]

$$J_0 = 4.5 \times 10^3, \quad \text{A/cm}^2\text{-}\mu\text{m}$$

$$\frac{g_0}{J_0} = 5 \times 10^{-2} \quad \text{cm-}\mu\text{m/A}$$

When the propagating TE_0 mode is not entirely within the active layer, the gain g gets modified as Γg, where Γ is the mode confinement factor. The mode confinement factor Γ is defined as the ratio of the light intensity, both within and outside the active layer. The value of Γ for the TE_0 mode of symmetric GaAs/Ga$_{1-x}$Al$_x$As double heterostructure laser with d between 0.01 and 2.0 μm at $x = 0.1$ to 0.6 has been analytically determined and is available elsewhere [6].

At low current density there is spontaneous emission in all directions. As the current density is increased, the gain increases until the threshold for lasing is reached, that is, the gain satisfies the condition that a light wave makes a complete round trip of the laser cavity without suffering any attenuation. This situation can be mathematically represented as

$$R \exp[(\Gamma g - \alpha) L] = 1 \tag{2.5}$$

where

Γ = confinement factor;
Γg = lasing threshold optical gain;
α = loss per unit length from free-carrier absorption and defect-center scattering;
L = length of the cavity;
R = reflectance of the ends of the cavity (if the reflectance of the ends are different, $R = \sqrt{R_1 R_2}$).

Equation 2.5 can be rewritten in terms of the threshold gain as [8]

$$\Gamma g = \alpha + \frac{1}{L} \ln\left(\frac{1}{R}\right) \tag{2.6}$$

As a numerical example, for GaAs, R is about 0.32 for uncoated facets and α is approximately 10/cm. This yields a threshold gain of 48/cm for L equal to 300 μm.

Equations (2.3) through (2.6) can be combined to yield [8]:

$$J_{th} = \frac{J_0 d}{\eta} + \frac{J_0 d}{\eta \Gamma g_0}\left[\alpha + \frac{1}{L}\ln\left(\frac{1}{R}\right)\right] \quad \text{A/cm}^2 \tag{2.7}$$

Therefore to reduce J_{th}, we can increase η, Γ, L, and R, and we can reduce d and α. As a numerical example, let us consider an Al$_x$Ga$_{1-x}$As laser diode with d equal to 0.05 μm and x equal to 0.3 for which Γ is 0.08 and η is unity. From equation (2.7), J_{th} is 825 A/cm^2 at room temperature. If x is increased to 0.6, Γ increases to 0.11 and hence J_{th} decreases to 661 A/cm^2. This decrease is because of the improvement in Γ with increasing x. Similar computations for a Ga$_x$In$_{1-x}$As$_y$P$_{1-y}$ laser show that for d equal to 0.2 μm, J_0 equal to 2400 A/cm^2-μm and g_0/J_0 equal to 0.057 cm-μm/A, J_{th} is about 1.5 kA/cm^2.

2.4 WAVEGUIDING PRINCIPLES AND INTENSITY PATTERN OF A DOUBLE HETEROSTRUCTURE LASER DIODE

The thin GaAs active layer of thickness $2d$, which is sandwiched between the two GaAlAs bonding layers, gives rise to a three-layered dielectric waveguide of the type illustrated in Figure 2.5. In this dielectric waveguide structure, the GaAlAs layer refractive index n_1 and n_3 are equal and less than the refractive index n_2 of the GaAs layer. The difference between the indices of refraction of GaAs and the GaAlAs with a molar fraction x is [6]:

$$\Delta n = n_2(\text{GaAs}) - n_1(\text{Ga}_{1-x}\text{Al}_x\text{As}) \approx 0.62\,x \tag{2.8}$$

Light rays propagate along this waveguide by total internal reflection at the interfaces. The rays interfere to produce a transverse standing wave pattern. The standing wave pattern propagates along the laser length unchanged in shape.

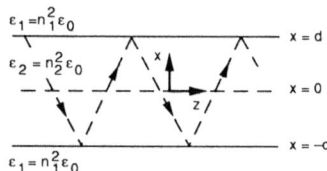

Figure 2.5 A symmetric slab waveguide.

In analyzing this structure, we consider the case of time harmonic behavior in the form of $\exp(i\omega t)$, where ω is the radian frequency, with infinite slab geometry so that there is no variation in the y direction; thus, $\partial/\partial y = 0$. Maxwell's equations for this geometry are

$$\nabla \times \mathbf{E} = -i\omega\mu\,\mathbf{H} \tag{2.9}$$

$$\nabla \times \mathbf{H} = i\omega\varepsilon\,\mathbf{E} \tag{2.10}$$

In these equations, μ and ε are the permeability and permittivity of the medium, respectively. Expanding these two equations, we obtain

$$\frac{\partial E_y}{\partial z} = i\omega\mu H_x \tag{2.11}$$

$$\frac{\partial E_x}{\partial z} - \frac{\partial E_z}{\partial x} = -i\omega\mu H_y \tag{2.12}$$

$$\frac{\partial E_y}{\partial x} = -i\omega\mu H_z \qquad (2.13)$$

$$\frac{\partial H_y}{\partial z} = -i\omega\varepsilon E_x \qquad (2.14)$$

$$\frac{\partial H_x}{\partial z} - \frac{\partial H_z}{\partial x} = i\omega\varepsilon E_y \qquad (2.15)$$

$$\frac{\partial H_y}{\partial x} = i\omega\varepsilon E_z \qquad (2.16)$$

Because the modes propagate in the z direction with the z dependence in the form of $\exp(-i\beta z)$, the derivative $\partial/\partial z$ is replaced by $-i\beta$, where β is the propagation constant in the z direction. Equations (2.11), (2.13), and (2.15) form a set that has a single electric field component, polarized parallel to the interface and perpendicular to the propagation direction, and hence is referred to as the *transverse electric* (TE) mode. With the above substitution, Maxwell's equations for the TE modes simplify as follows:

$$E_y = -\frac{\omega\mu}{\beta} H_x \qquad (2.17)$$

$$\frac{\partial E_y}{\partial x} = -i\omega\mu H_z \qquad (2.18)$$

Equations (2.12), (2.14), and (2.16) form a set which have a single magnetic field component polarized parallel to the interface and perpendicular to the propagation direction, and therefore is referred to as *transverse magnetic* (TM) modes. As in the previous case Maxwell's equations for the TM modes simplify as follows

$$H_y = \frac{\omega\varepsilon}{\beta} E_x \qquad (2.19)$$

$$E_z = \frac{-i}{\omega\varepsilon} \frac{\partial H_y}{\partial x} \qquad (2.20)$$

Insofar as the waveguide is symmetric about the plane $x = 0$, the mode solutions must be either even or odd in x:

$$E_y(x,z,t) = E_y(-x,z,t) \qquad (2.21)$$

in the case of even modes, and

$$E_y(x,z,t) = -E_y(-x,z,t) \tag{2.22}$$

in the case of odd modes.

2.4.1 Even TE Modes

The solution for the even mode is in the form:

$$E_y = A \exp[-k_1(|x| - d)] \exp(-i\beta z), \qquad |x| \geq d \tag{2.23}$$

as the fields decay exponentially outside the waveguide, and

$$E_y = B \cos(k_2 x) \exp(-i\beta z), \qquad |x| \leq d \tag{2.24}$$

as a standing wave is set up inside the waveguide. In these expressions, k_1 and k_2 are positive, real, x-directed propagation constants to be determined. From equation (2.18), we obtain

$$H_z = \mp i \frac{k_1 A}{\omega\mu} \exp[-k_1(|x| - d) - i\beta z], \qquad |x| \geq d \tag{2.25}$$

where the negative sign is used with $\geq d$ and the positive sign is used with $\leq -d$, and

$$H_z = -i \frac{k_2 B}{\omega\mu} \sin(k_2 x) \exp(-i\beta z), \qquad |x| \leq d \tag{2.26}$$

The tangential field components E_y and H_z are continuous across the interface. Applying the continuity condition to equations (2.23) and (2.24) at $x = \pm d$ yields

$$A = B \cos(k_2 d) \tag{2.27}$$

Similarly, from equations (2.25) and (2.26), we obtain

$$k_1 A = k_2 B \sin(k_2 d) \tag{2.28}$$

From equations 2.27 and 2.28 we obtain

$$k_1 d = k_2 d \tan(k_2 d) \tag{2.29}$$

The solutions for E_y in equations (2.23) and (2.24) must also satisfy the wave equations:

$$\frac{\partial^2}{\partial x^2} E(x,y) + (k_0^2 n_1^2 - \beta^2)E(x,y) = 0 \qquad (2.30)$$

and

$$\frac{\partial^2}{\partial x^2} E(x,y) + (k_0^2 n_2^2 - \beta^2)E(x,y) = 0 \qquad (2.31)$$

for $|x| \geq d$ and $|x| \leq d$, respectively, where k_0 is the free-space propagation constant and is equal to ω/c. These equations result in

$$\beta^2 = k_0^2 n_1^2 + k_1^2 \qquad (2.32)$$

$$\beta^2 = k_0^2 n_2^2 - k_2^2 \qquad (2.33)$$

By combining the above two equations we obtain

$$(k_2 d)^2 + (k_1 d)^2 = (n_2^2 - n_1^2) k_0^2 d^2 \qquad (2.34)$$

The propagation constants k_1 and k_2 of a given mode must simultaneously satisfy equations (2.29) and (2.34). A graphical solution is illustrated in Figure 2.6,

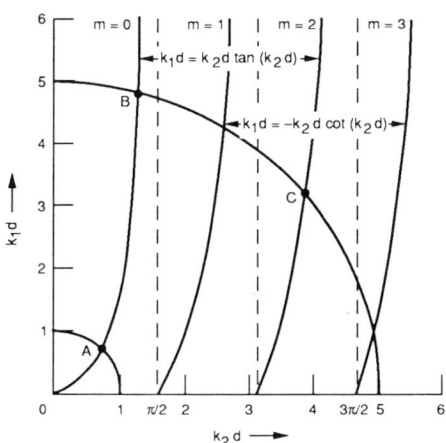

Figure 2.6 Graphical solution of eigenvalue equation of slab waveguide.

and consists of finding the intersections in the k_1d–k_2d plane of the circle defined by

$$(k_2d)^2 + (k_1d)^2 = (n_2^2 - n_1^2)k_0^2 d^2 = u^2 \tag{2.35}$$

with the curve defined by

$$k_1d = k_2d \tan(k_2d) \tag{2.36}$$

Each intersection with a $k_1 > 0$ corresponds to a confined mode. The number of possible solutions and the number of transverse modes increase with the slab thickness $2d$ and the index difference. To guarantee operation of the laser in the fundamental transverse mode, the index difference and the slab thickness must be small. The propagation constant β of a given mode can be obtained from equations (2.32) and (2.33), once k_1 and k_2 are determined. Intensity distributions corresponding to the three lowest order modes are shown in Figure 2.7. For small values of u such that

$$0 < u < \pi \tag{2.37}$$

only one intersection (point A) exists between the circle and the curve. This mode is designated as the TE_0 mode and has a k_2 value within the range of

$$0 < k_2d < \pi/2 \tag{2.38}$$

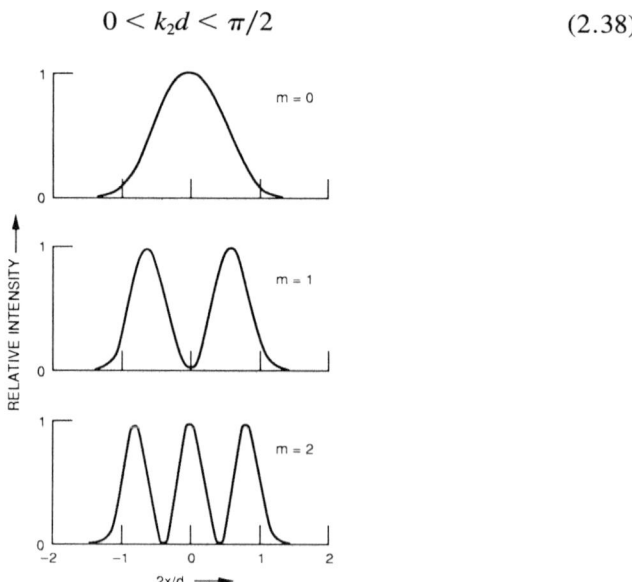

Figure 2.7 Intensity distribution of a symmetric slab waveguide.

and has a zero crossing in the interior of the slab $|x| \leq d$. When u takes a value between

$$\pi < u < 2\pi \tag{2.39}$$

we obtain two intersections. The first intersection (point B) corresponds to a value of k_2, which is such that $k_2 d < \pi/2$, and also corresponds to the lowest order TE_0 mode. The second intersection (point C) corresponds to a value of k_2 such that

$$\pi < k_2 d < 3\pi/2 \tag{2.40}$$

and consequently has two zero crossings in the region $|x| < d$. This is the TE_2 mode. Both TE_0 and TE_2 correspond to the same frequency and thus can be simultaneously excited by the same input field. Further, the TE_0 mode has a higher value of k_1 than the TE_2 mode, which suggests that the TE_0 mode is more highly confined to the interior of the slab. Thus, the effective interaction between the carriers and the photons is greater for the TE_0 mode, thereby resulting in a higher gain. From equations (2.32) and (2.33), the propagation constant of the TE_0 mode hence is greater than that of the TE_2 mode so that the phase velocity of the TE_0 mode is less than that of the TE_2 mode.

2.4.2 Odd TE Modes

The odd mode can be described by the following set of equations:

$$E_y = A \exp[-k_1(|x| - d) - i\beta z] \qquad |x| \geq d \tag{2.41}$$

$$E_y = B \sin(k_2 x) \exp(-i\beta z) \qquad |x| \leq d \tag{2.42}$$

Proceeding in a manner similar to that of the even mode case leads to

$$k_1 d = -k_2 d \cot(k_2 d) \tag{2.43}$$

The lowest order odd TE mode is designated as TE_1 because k_2 satisfies

$$\pi/2 < k_2 d < \pi \tag{2.44}$$

The propagation constant of the TE_1 mode lies between the TE_0 and TE_2 modes.

2.4.3 TM Modes

The characteristics of the TM modes are similar to that of the TE modes. However, the value of k_1 is observed to be less than the corresponding k_1 for the TE case, implying that a larger fraction of the mode power propagates outside the active region, which is undesirable.

2.4.4 Effect of Facet Reflectivity on Mode Selection

The light emitted from a double heterostructure laser has been observed to be predominantly TE rather than TM. Also, the fundamental TE_0 mode is often the only mode observed at active layer thicknesses where higher order modes are possible [9].

In equation (2.6), we indicated that the gain necessary to reach threshold depends on both the confinement factor and the facet reflectivity. As the values of k_1 and k_2 are about the same for TE and TM modes, the confinement factor Γ does not significantly differ for TE and TM modes. This implies that mode selection is due to the difference in facet reflectivity for TE and TM modes. Rigorous analysis [10] of facet reflectivity does not result in closed form solutions, and so the ensuing numerical results will only be qualitatively described. In these results, the laser wavelength is taken as 0.86 μm, and the refractive index n_1 and n_2 are taken as 1.0 and 3.60, respectively. In Figure 2.8, the computed reflection coefficient for the dom-

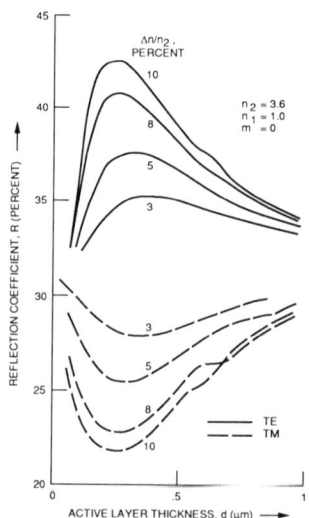

Figure 2.8 Computed reflectivity for the fundamental mode as a function of active layer thickness.
Source: Ikegami, T., "Reflectivity of Mode at Facet and Oscillation Mode in Double-Heterostructure Injection Lasers," *IEEE J. Quantum Electronics*, Vol. QE-8, No. 6, June 1972, pp. 470–476. Reprinted with permission.

inant TE and TM modes is presented as a function of the active layer thickness. The parameter for the various curves is the refractive index step, given as

$$\frac{\Delta n}{n_2} = \frac{n_2 - n_1}{n_2} \qquad (2.45)$$

and expressed as a percentage. The continuous curves are for TE_0 mode, and the broken curves are for TM_0 mode. This result demonstrates that the larger reflectance for the TE_0 mode is responsible for the oscillations in a double heterostructure laser diode to be predominantly TE.

In Figure 2.9, the computed mode loss for the dominant as well as the higher order TE and TM modes, is presented as a function of the active layer thickness. The parameters for the various curves, is the order of the mode. The ratio $\Delta n/n_2$ is held fixed at 5% which corresponds to an AlAs mole fraction of approximately 0.25 used in a practical laser diode. This result demonstrates that the loss decreases as the order of the TE mode increases, and is therefore a second reason for the laser oscillations to be predominantly TE. Therefore, we can conclude that for a practical GaAs/GaAlAs laser diodes the active layer thickness should be on the order of 0.3 μm for lower loss, and enhanced interaction between the carriers and the photons. This condition also ensures that the laser cavity oscillates in the fundamental TE_0 mode.

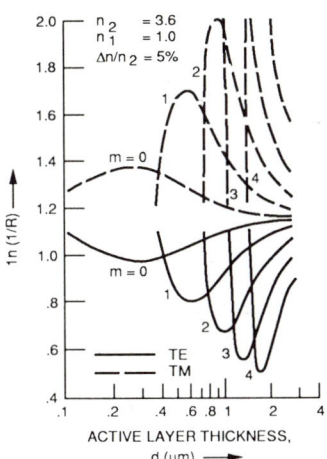

Figure 2.9 Computed mode loss as a function of active layer thickness.
  Source: Ikegami, T., "Reflectivity of Mode at Facet and Oscillation Mode in Double-Heterostructure Injection Lasers," *IEEE J. Quantum Electronics*, Vol. QE-8, No. 6, June 1972, pp. 470–476. Reprinted with permission.

2.5 FUNDAMENTAL DEVICE CHARACTERISTICS

2.5.1 Light Intensity *versus* Current

When the bias current is increased from low current densities to high current densities in excess of the threshold, the light output from the laser increases and also changes from spontaneous emission, to stimulated emission. Figure 2.10 shows a typical light intensity *versus* current characteristics of a double heterostructure laser. The extrapolation of this curve to zero optical power gives the threshold current, I_{th}. The external differential quantum efficiency η_d is defined as the number of photons emitted per radiative electron-hole pair recombination above threshold, that is

$$\eta_d = \frac{d(P/h\nu)}{d[(I - I_{th})/q]}$$

$$= \frac{q}{E_g} \frac{dP}{dI} \qquad (2.46)$$

where

dP = incremental change in the emitted optical power;
dI = incremental change in the drive current;
$h\nu \approx E_g$ = band gap energy in eV;
q = electron charge in C.

Thus, η_d is calculated from the straight line portion of the curve for the emitted optical power *versus* the drive current. Figure 2.10 also shows the effect of ambient temperature on the laser diode characteristics. It has been experimentally observed

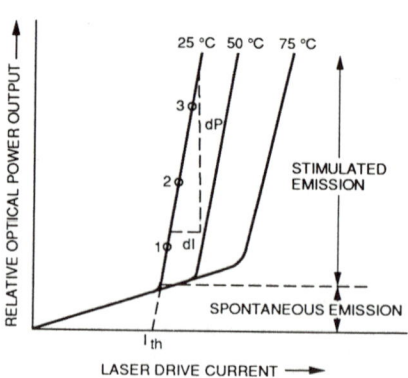

Figure 2.10 Light output *versus* the laser diode current.

that the threshold current increases exponentially with temperature as [14]

$$I_{th} \approx \exp(T/110). \tag{2.47}$$

where T is the ambient (heat sink) temperature in °C.

2.5.2 Optical Spectra

If L is the length of the laser cavity between reflection planes and n is the refractive index of the cavity material at a wavelength λ, only longitudinal modes that satisfy the following relation exist:

$$m\left(\frac{\lambda}{2n}\right) = L \tag{2.48}$$

where m is the integral number of half wavelengths. The separation $\Delta\lambda$ between these allowed longitudinal modes is the difference in the wavelength corresponding to m and $m + 1$. Differentiating equation (2.48) with respect to λ, we obtain

$$\Delta\lambda = \frac{\lambda^2 \Delta m}{2nL[1 - (\lambda/n)(dn/d\lambda)]} \tag{2.49}$$

for large m.

Figure 2.11 shows the spectral distribution for a typical double heterostructure

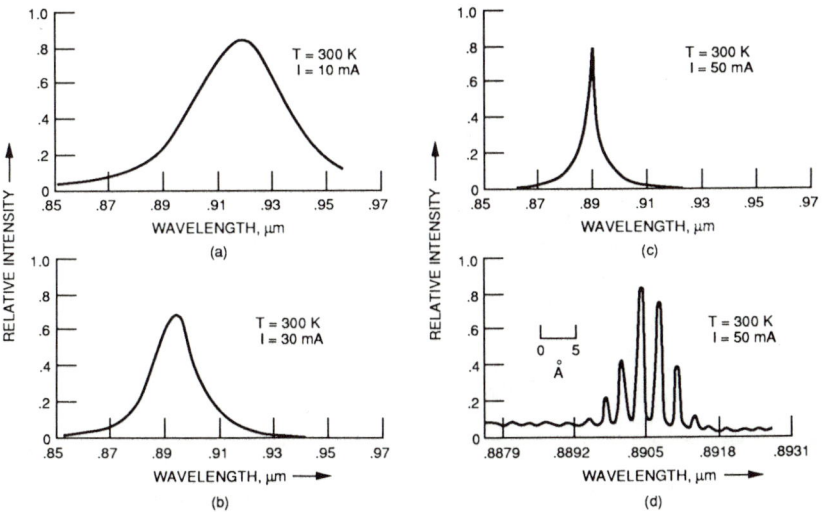

Figure 2.11 Typical spectral distribution for a double heterostructure laser diode.

laser diode. At low currents, the spectral distribution has a broad half-power width typically several hundred angstroms wide (Figure 2.11(a)). As the current approaches the threshold current, the spectral width becomes narrower (Figure 2.11(b)). If above threshold when lasing is initiated, the emission spectra narrows considerably (Figure 2.11(c)). Figure 2.11(d) shows the spectrum of Figure 2.11(c) under high resolution. The emission lines that are evenly spaced with a separation $\Delta\lambda$ belong to the longitudinal modes that we discussed above.

2.5.3 Near- and Far-Field Patterns

Figure 2.12 shows the computed [11] near- and far-field intensity distribution for several active region thicknesses, at a fixed refractive index step. Figure 2.12(a)

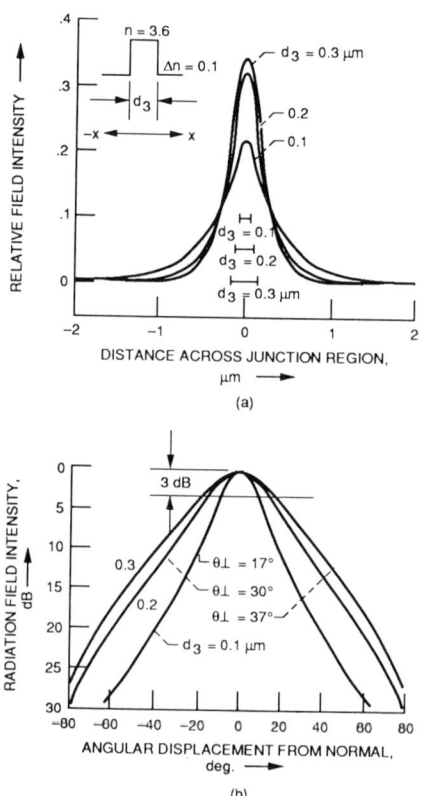

Figure 2.12 Computed near- and far-field pattern of a double-heterostructure laser.
    Source: Butler, J.K., and H. Kressel, "Design Curves for Double-Heterojunction Laser Diodes," RCA Review, Vol. 38, December 1977, pp. 542–558. Reprinted with permission.

shows that as the active layer thickness increases there is an increase in the confinement which results in a sharper near-field intensity pattern. Further, as better confinement results in a narrow source, the far-field intensity pattern (radiation pattern) broadens with concommitant increase in the full angle at half power (3 dB beamwidth), as shown in Figure 2.12(b). Figure 2.13 gives a schematic representation of the far-field emission of a double heterostructure laser diode [12]. The full angles at half-power are $\theta_\perp$ and $\theta_\parallel$ in the directions perpendicular to and along the junction plane, respectively. If the physical geometry of the laser is known, $\theta_\perp$ can be determined from the following closed-form expression [13]:

$$\theta_\perp = 2 \tan^{-1}\left[\frac{0.59 \lambda}{w_0 \pi}\right] \quad \text{(degrees) for } 1.5 < D < 6 \qquad (2.50)$$

where

$w_0 = T[0.31 + (3.15/D^{1.5}) + (2/D^6)]$
$T = 2d =$ Active layer thickness
$\lambda =$ free space wavelength.

$$\theta_\perp = \frac{0.65 \, D\sqrt{n_2^2 - n_1^2}}{1 + 0.086 \, sD^2} \quad \text{(degrees) for } D < 1.5 \qquad (2.51a)$$

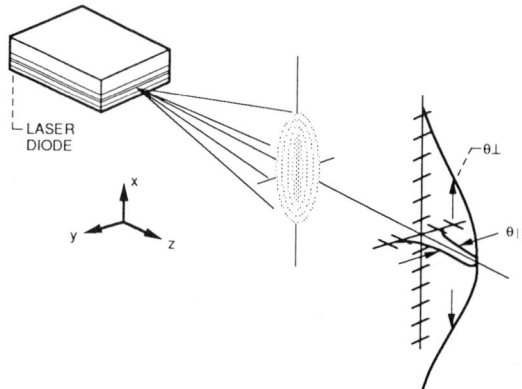

Figure 2.13 Schematic representation of far-field emission of a double-heterostructure laser diode.

where

$$D = \frac{2\pi T}{\lambda}\sqrt{n_2^2 - n_1^2}$$

$$s = \frac{2.52\sqrt{n_2^2 - n_1^2}}{\tan^{-1}(0.36\sqrt{n_2^2 - n_1^2})} - 5.17$$

Because the TE_0 mode has a single electric field component polarized parallel to the interface without variation in the y direction, $\theta_\parallel$ can be approximated as [37]

$$\theta_\parallel \approx 0.886 \frac{\lambda}{w_1} \frac{180}{\pi} \quad \text{(degrees)} \qquad (2.51b)$$

where w_1 is the extent of wave confinement in the y direction. Typically, $\theta_\parallel$ is on the order of 8 to 10°, whereas $\theta_\perp$ is considerably larger and typically varies between 35 to 65 degrees, depending on the active layer thickness [14].

2.6 LASER DIODE TO A SINGLE-MODE OPTICAL FIBER COUPLING

In a practical system, to reduce the loss when coupling an optical source, such as a laser diode to a single-mode fiber, the coupling efficiency is important. In a typical stripe geometry laser, the dimensions of the radiating aperture are about 0.5 μm × 5 μm, causing the radiating beam to be elliptical. Hence, the angular width of the output beam in the direction perpendicular to the thin junction of thickness 0.5 μm is about 55°, whereas, in the orthogonal direction, the width is about 5 μm and the divergence angle is about 10°. Therefore, in order to couple the output radiation into a single-mode fiber we must focus the beam in the plane of large divergence.

Figure 2.14 illustrates several methods [15–18] of coupling between a semiconductor laser and a single-mode fiber. The measured coupling efficiency in each case is summarized in Table 2.1. In order to make a valid comparison the laser diode, optical fiber, and operating wavelength are about the same in all the cases.

2.7 DIRECT CURRENT MODULATION OF SEMICONDUCTOR LASERS

In Section 2.5.1, we have seen that the light output from a semiconductor laser diode varies with the change of injection current (Figure 2.10). Thus, by injecting the modulating microwave signal along with the dc bias, we can directly intensity-modulate the laser diode output. The setup for carrying out this experiment is explained later in the chapter on RF fiber optic links. However, in this Section, we shall explore

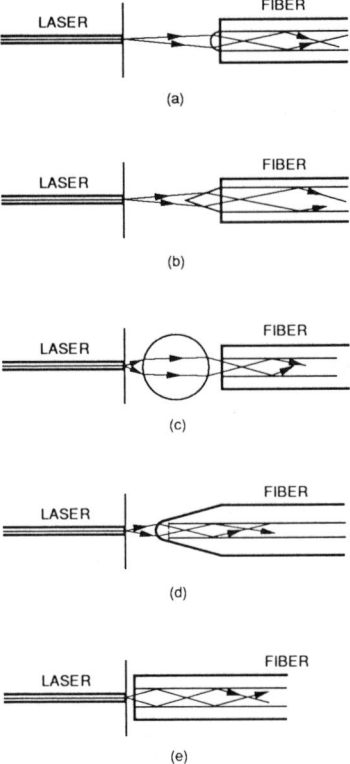

Figure 2.14 Techniques to couple a semiconductor laser diode to a single-mode optical fiber: (a) hemispherical microlens; (b) conical lens; (c) spherical lens; (d) tapered fiber with high-index lens; (e) direct coupling (butt coupling).

the theoretical limit for the maximum modulation frequency and the techniques for increasing this limit.

We can explain the laser dynamics in the most basic manner by a pair of rate equations, which are functions of the photon and carrier densities inside the laser medium [19], [20] as given below.

$$\frac{dN}{dt} = \frac{I}{qV} - \frac{N}{\tau} - A(N - N_{tr})\,P \tag{2.52}$$

$$\frac{dP}{dt} = A(N - N_{tr})\,P\,\Gamma - \frac{P}{\tau_p} \tag{2.53}$$

Table 2.1
Coupling Efficiencies between Lasers and Single-Mode Optical Fiber

Laser Type and Wavelength	Fiber Type and Core Diameter	Method of Coupling	Coupling Loss (dB)	Reference
(InGaAsP Buried-Heterostructure, 1.3 μm)	Single-Mode, 8 μm	Hemispherical Microlens	2.9	[15]
	Single-Mode, 8 μm	Conical Lens	3.0	[16]
	Single-Mode, 10 μm	Spherical Lens	4.0	[17]
	Single-Mode,	Tapered Fiber with High-Index Lens	2.6	[18]
	Single-Mode, 8 μm	Butt Coupling	>10.0	[16]

where

- N = injected electron (and hole) density;
- P = photon density inside the active region of a semiconductor laser diode;
- I = total current;
- V = volume of the active region;
- τ = spontaneous recombination lifetime;
- τ_p = photon lifetime as limited by the absorption in the bounding medium, scattering, and coupling through the output mirrors;
- Γ = confinement factor;
- N_{tr} = inversion density needed to achieve transparency (i.e., the electron density in the medium must exceed a certain level to exhibit positive gain);
- A = optical gain coefficient.

Equation (2.52) states that the rate of increase in carrier density is equal to the rate of current injection (I/qV), less the rate of carrier loss due to spontaneous recombination ($-N/\tau$), less the loss of carriers due to stimulated recombination [$-A(N - N_{tr}) P$]. Equation (2.53) states that the rate of increase in photon density is equal to the rate of photon generation by stimulated emission [$A(N - N_{tr}) P \Gamma$], less the loss of photons due to cavity dissipation ($-P/\tau_p$). The contribution of spontaneous emission to the photon density is neglected since only a very small fraction of the spontaneously emitted power enters the lasing mode.

The steady-state solutions (dc solutions) N_0 and P_0 are obtained by equating the left-hand side of (2.52) and (2.53) to zero; thus,

$$0 = \frac{I_0}{qV} - \frac{N_0}{\tau} - A(N_0 - N_{tr}) P_0 \qquad (2.54)$$

$$0 = A(N_0 - N_{tr}) P_0 \Gamma - \frac{P_0}{\tau_p} \qquad (2.55)$$

Let us assume that the current is composed of a dc and ac component

$$I = I_0 + i_1 \exp(i\omega t) \qquad (2.56)$$

and define the small-signal modulation response n_1 and p_1 by

$$N = N_0 + n_1 \exp(i\omega t) \qquad (2.57)$$

$$P = P_0 + p_1 \exp(i\omega t) \qquad (2.58)$$

Using (2.55), (2.56), (2.57) and (2.58) in (2.52) and (2.53) yields the following small-signal equations:

$$-i\omega n_1 = -\frac{i_1}{qV} + \left(\frac{1}{\tau} + A P_0\right) n_1 + \frac{1}{\Gamma} \frac{p_1}{\tau_p} \qquad (2.59)$$

$$i\omega p_1 = A P_0 \Gamma n_1 \qquad (2.60)$$

Solving equations (2.59) and (2.60) for the modulation response $p_1(\omega)/i_1(\omega)$, we obtain

$$p_1(\omega) = \frac{-(i_1/qV) A P_0 \Gamma}{\omega^2 - i\omega/\tau - i\omega A P_0 - A P_0/\tau_p} \qquad (2.61)$$

In the following section, we will discuss a method to measure the modulation response.

2.7.1 Measurement of Small Signal Modulation Response of a Laser Diode

Figure 2.15 illustrates an experimental setup [21], which consists of a microwave sweep oscillator, scattering parameter test set, and network analyzer. The microwave

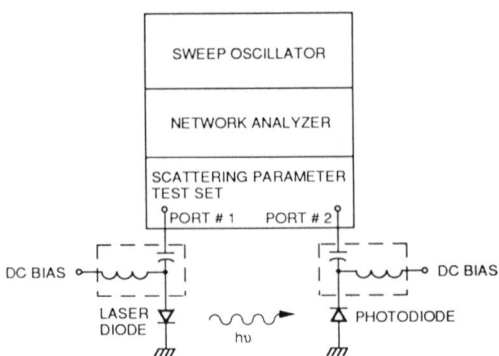

Figure 2.15 Experimental setup for measuring the frequency response of a laser diode.

signal from port 1 of the network analyzer is coupled to a laser diode through a bias tee. The intensity-modulated optical signal is detected by a photodiode. The output of the photodiode is a microwave signal coupled through a bias tee to port 2 of the network analyzer. Figure 2.16 shows a typical measured normalized-amplitude response of a laser diode as a function of the modulating frequency for several bias currents which are indicated in Figure 2.10. The measured response is typically observed to be flat at low frequencies, but has a resonant peak just before it drops into a 40 dB/decade attenuation. The frequency corresponding to the resonant peak is known as the relaxation frequency and denoted as ω_r. For precise measurements, the 3 dB bandwidth of the photodiode should be much higher than ω_r.

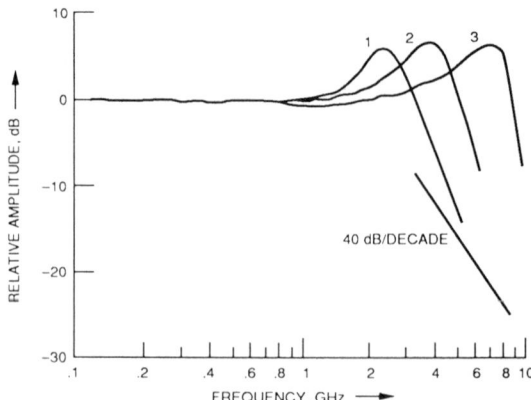

Figure 2.16 Typical frequency response of a laser diode.

2.7.2 Expression for the Relaxation Resonance Frequency

An expression for the peak frequency is obtained by minimizing the magnitude of the denominator in equation (2.61); thus,

$$\omega^2 - i\omega/\tau - i\omega A P_0 - A P_0/\tau_p = 0 \qquad (2.62)$$

which results in

$$\omega_r = \sqrt{\frac{A P_0}{\tau_p} - \frac{1}{2}\left[\frac{1}{\tau} + A P_0\right]^2} \qquad (2.63)$$

The contribution from the second term under the square root sign is usually small and hence can be neglected; therefore,

$$\omega_r = \sqrt{\frac{A P_0}{\tau_p}} \qquad (2.64)$$

This result suggests that increasing ω_r, and thus increasing the input linear region of the modulated response $p_1(\omega)/i_1(\omega)$, the optical gain coefficient A should increase, the photon lifetime τ_p should decrease, and the laser should operate at an internal photon density P_0 as high as possible.

The gain coefficient A can be increased roughly by a factor of 5, by cooling the laser from room temperature to liquid nitrogen temperature of 77 K, [7,22]. Biasing the laser at higher currents would increase the photon density P_0 in the active region, which simultaneously increases the optical output power density. However, catastrophic mirror damage occurs at a power density of about 1 MW/cm^2, for a laser with a mirror reflectivity of 0.3 [22,23]. This sets an upper limit on the maximum permissible photon density, and hence the maximum modulation bandwidth. This limit can be increased by incorporating a window structure [24]. The third way to increase the modulation bandwidth is to reduce the photon lifetime τ_p by decreasing the length of the laser cavity. However, such a laser has to operate at a high current density, and therefore, thermal effects due to heating will limit the maximum attainable modulation bandwidth [22]. As an example, a laser with a cavity length of 300 μm operating at an output optical power density of 0.8 MW/cm^2 possesses a bandwidth of 5.5 GHz, and corresponding pump current density is 3 kA/cm^2. Reducing the cavity length to 100 μm and operating at an identical power level results in a bandwidth of 8 GHz, but the corresponding current density is 6 kA/cm^2 [22]. The thermal effects alone may cause the laser to degrade rapidly.

Finally, equation (2.64) for ω_r can be recast as follows:

$$\omega_r = \sqrt{\frac{1 + A\tau_p \Gamma N_{tr}}{\tau \cdot \tau_p}\left(\frac{I_0}{I_{th}} - 1\right)} \qquad (2.65)$$

We can easily see that there is a direct correspondence between the above equation and Figure 2.16, which shows that ω_r increases as I_0 increases (Figure 2.10).

2.8 SMALL-SIGNAL LUMPED ELEMENT EQUIVALENT CIRCUIT MODEL

A small-signal lumped element equivalent circuit model [25] for a typical packaged laser diode is shown in Figure 2.17. The series R_s accounts for *contact resistance* and *bulk resistance* of the semiconductor chip; C_s represents *parasitic capacitance* associated with the chip; C_{sc} is the *space charge capacitance* and C_d is the *active layer diffusion capacitance*. The current-controlled voltage source provides a voltage v_0 at the output port of the model and is an analog of the small-signal photon density in the laser. The output voltage v_0 can therefore be used as a measure of the small signal light output intensity from the laser.

The elements of the equivalent circuit is determined by an optimization software such as EEsof's Touchstone® [26]. The optimization routine adjusts the element values so as to improve the match between the measured and calculated reflection coefficient and the transmission coefficient at a number of dc bias conditions, above and below threshold.

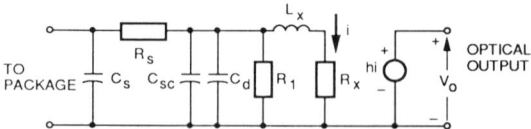

Figure 2.17 Small-signal lumped-element equivalent circuit model of a laser diode.
Source: After Tucker, R.S., and D.J. Pope, "Microwave Circuit Models of Semiconductor Injection Lasers," *IEEE Trans. Microwave Theory Tech.*, Vol. MTT-31, No. 3, March 1983, pp. 289–294.

2.9 ADVANCED HETEROSTRUCTURE LASERS

2.9.1 GaAs/GaAlAs Buried Heterostructure Laser

The double heterostructure laser (Figure 2.3) discussed in Section 2.3 is incapable of confining the current and the radiation in the y direction. As a consequence, (1) the laser can support more than one transverse mode resulting in mode hopping; (2) the far field beam widths, $\theta_\perp$ and $\theta_\parallel$, are inequal and therefore require special lens for coupling to a single-mode device, such as, an optical fiber. These difficulties are overcome if some transverse optical and carrier confinement are provided. An excellent example of this approach is the buried heterostructure laser [27] shown in Figure 2.18.

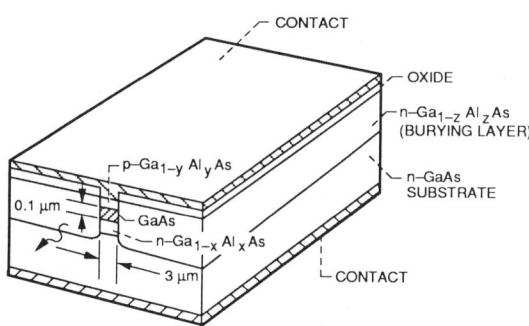

Figure 2.18 A typical GaAs-GaAlAs buried heterostructure laser.
Source: After Tsukada, T., "GaAs-GA$_{1-x}$As Buried-Heterostructure Injection Lasers," *J. Applied Physics*, Vol. 45, No. 11, November 1974, pp. 4899–4906.

In this laser, the three layers—the *n*-type Ga$_{1-x}$Al$_x$As cladding layer, the undoped GaAs active layer, and the *p*-type Ga$_{1-y}$Al$_y$As cladding layer—are sequentially grown on an *n*-type GaAs crystalline substrate. The structure is etched through a mask down to the substrate leaving a thin rectangular mesa composed of the original layers. A Ga$_{1-z}$Al$_z$As burying layer is grown on both sides of the mesa.

In the resulting structure, the GaAs active layer is surrounded by a lower index GaAlAs region so that the electromagnetic structure is that of a rectangular dielectric waveguide. The transverse dimensions and the refractive index discontinuities for this waveguide are chosen so that only the lowest order transverse mode can propagate. Here, we mention that laser structures which employ a variation in the real refractive index along the junction plane to form an optical waveguide are referred to as *index guided lasers* [6]. Another interesting feature of this structure is that the discontinuities in the energy band structure at the GaAs/GaAlAs interface help to confine the injected carriers to the active region. The discontinuities act as potential barriers to prevent the escape of carriers out of the active region. The difference of beamwidth in the two orthogonal planes is expected to be small owing to the reduction in the width of the active region.

Commercially available GaAs/GaAlAs buried heterostructure lasers [21,28] operate at a peak wavelength of about 0.84 μm at room temperature. The threshold current is typically about 20 mA. The beamwidths $\theta_\perp$ and $\theta_\parallel$ are typically about 35 and 25°, respectively, and the 3 dB bandwidth of the laser is about 10 GHz.

2.9.2 InP/InGaAsP Buried Heterostructure Laser

An InP/InGaAsP planar buried heterostructure laser [29] is schematically illustrated in Figure 2.19. The mesa of the laser consists of four layers: an n-type InP cladding layer; an undoped InGaAsP active layer; a p-type InP cladding layer; a p-type cap layer. A semi-insulating InP burying layer is then grown on both sides of the mesa. The semi-insulating region also helps to reduce the parasitic capacitance.

The laser operates at a peak wavelength of 1.31 μm at room temperature. The laser has a low threshold current, typically about 17 mA and has a high external differential quantum efficiency of about 42%. The beamwidths $\theta_\perp$ and $\theta_\parallel$ are about 32 and 22°, respectively, and are comparable to the GaAs laser discussed in the previous section. The CW power output of the laser is about 8 to 12 mW at room temperature and is adequate for communication applications. Because of the small parasitic capacitance, the laser has a wide bandwidth, which spans up to 10 GHz at room temperature.

The primary limitation to the bandwidth of most laser structures is the parasitic capacitance associated with the bonding pad and the structure used for current confinement. One of the successful techniques that has been used to reduce the parasitic capacitance is to use a thick polyimide layer under the bonding pad, and to use a constricted-mesa design for current confinement [30]. Figure 2.20 illustrates the laser structure which incorporates this technique. The measured bandwidth of the laser is about 16 GHz at room temperature. The disadvantage of the above technique is that the polyimide layers and the constricted mesa increase the thermal resistance. This is evident from the observed increase in the bandwidth of the laser to 26.5 GHz, by cooling it to $-60°$ C.

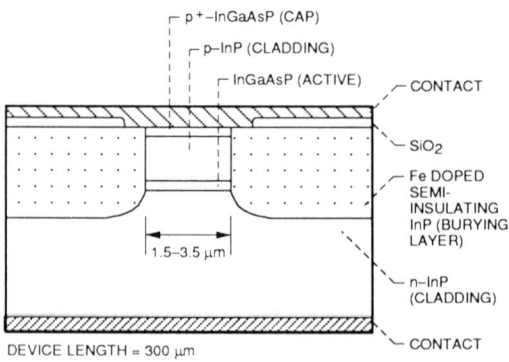

Figure 2.19 A typical InP/InGaAsP buried-heterostructure laser.
Source: After Wakao, K., K. Nakai, T. Sanada, M. Kuno, T. Odagawa, and S. Yamakoshi, "InGaAsP/InP Planar Buried Heterostructure Lasers with Semi-Insulating InP Current Blocking Layers Grown by MOCVD," *IEEE J. Quantum Electronics*, Vol. QE-23, No. 6, June 1987, pp. 943–947.

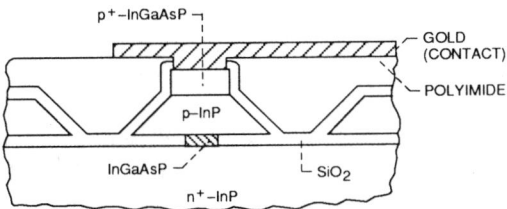

Figure 2.20 A typical InP/InGaAsP constricted-mesa laser.
Source: After Bowers, J.E., B.R. Hemenway, T.J. Bridges, E.G. Burkhardt, and D.P. Wilt, "26.5 GHz Bandwidth InGaAsP Lasers with Tight Optical Confinement," *Electronics Letters*, Vol. 21, No. 23, November 1985, pp. 1090–1091.

Lastly, for applications which require both high power and wide bandwidth, buried crescent lasers [31] are recommended. Figure 2.21 shows the cross section of a buried crescent laser. Typical cw power output from the laser is about 30 mW per facet at 20° C. The laser is capable of being directly modulated at a frequency as high as 8.3 GHz.

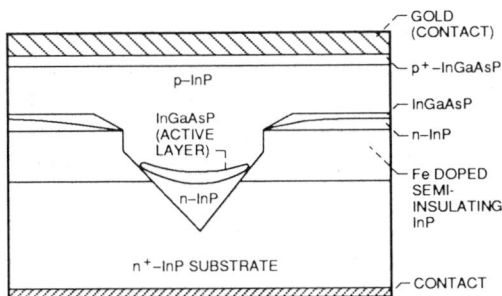

Figure 2.21 A typical InP/InGaAsP buried crescent laser.
Source: After Zah, C.E., J.S. Osinski, S.G. Menocal, N. Tabatabaie, T.P. Lee, A.G. Dentai, and C.A. Burrus, "Wide-Bandwidth and High-Power 1.3 μm InGaAsP Buried-Crescent Lasers with Semi-Insulating Fe-Doped InP Current Blocking Layers," *Electronics Letters*, Vol. 23, No. 1, January 1987, pp. 52–53.

2.9.3 InP/InGaAsP Distributed Feedback Laser

In the laser-diode structures described so far the radiation is generated within a Fabry-Perot resonator cavity. The two end walls of this rectangular resonant cavity are formed by mirror facets. The mirror facets are constructed by making two parallel cleaves along natural cleavage planes of the semiconductor crystal. In an optical integrated circuit, in which the laser diodes are monolithically integrated within the semiconductor wafer, the mirror facets result in the disruption of the wafer surface. Hence, an alternative approach would be to incorporate a Bragg type of diffraction grating into the multilayer structure of the laser diode along its length. The grating is usually produced by corrugating the interface between two of the semiconductor layers that compose the laser diode. The corrugation provides 180 degrees reflection at a free-space wavelength, λ, given by

$$\lambda = \frac{2 s n_g}{1} \qquad (2.66)$$

where

s = grating spacing;
1 = order of the grating and is equal to 1, 2, 3 ...;
n_g = effective refractive index in the waveguide for the mode under consideration.

As the corrugation provides distributed feedback for light in a very narrow spectral range, the lasing occurs with a pure single frequency. In a discrete laser, the full optical output is needed only from the front facet of the laser. In this case, a dielectric reflector is deposited on the rear laser facet, which not only reduces the optical loss in the cavity, but also reduces the threshold current and increases the external quantum efficiency. In addition, the light-emitting facet is processed with antireflective coating such as, Si_3N_4 to reduce mirror reflection so that the Fabry-Perot mode of oscillation is suppressed.

Figure 2.22 shows the schematic cross section of a mesa-fabricated, buried-heterostructure *distributed feedback* (DFB) laser diode [32]. The laser operates at a peak wavelength of 1.3 μm at room temperature. The parasitic capacitance of the laser is greatly reduced by forming two channels on either side of the strip. The

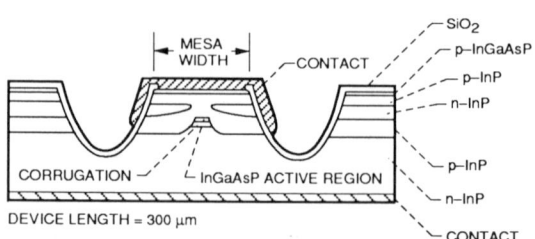

Figure 2.22 A typical InP/InGaAsP distributed feedback laser.
Source: After Kamite, K., H. Sudo, M. Yano, H. Ishikawa and H. Imai, "Ultra-High Speed InGaAsP/InP DFB Lasers Emitting at 1.3 μm Wavelength," *IEEE J. Quantum Electronics*, Vol. QE-23, No. 6, June 1987, pp. 1054–1058.

reduction in parasitic capacitance resulted in a bandwidth of 13.9 GHz which is considered to be the highest for a DFB laser [33]. By controlling the facet reflectivity through appropriate choice of Si_3N_4 film thickness, the lowest reported relative intensity noise of -160 dB/Hz was obtained [34]. The spectral width is typically as small as 0.25 to 0.4 nm under modulation [33]. This is about one-tenth of that of Fabry-Perot laser diodes discussed in Sections 2.9.1 and 2.9.2. In addition, a wavelength stability against temperature of about 0.9 angstrom/°C is attained owing to the diodes grating structure [33]. Lastly, no appreciable change in the driving current was observed when the laser was operated at 70° C at 8 mW per facet CW optical power output for 1600 hours. The reliability of the laser is therefore considered to be good [32,33]. These unique features of the laser favor high speed microwave communication applications.

2.10 RELIABILITY OF LASER DIODES

The lifetime of 0.8 μm GaAs/GaAlAs laser diodes at low temperature and low power is on the order of 10^6 hours [35]. We also advise you to refer to the Chapter on optoelectronic switch matrix for greater details. However, the lifetime of 1.3 μm InP/InGaAsP buried-heterostructure laser diodes under constant power of 5 mW/facet at about 10° C, is 10^7 hours [36]. Lastly, in the case of distributed feedback lasers, the wavelength shifts to the shorter wavelength with aging. This is because the reduction in refractive index due to the increase in threshold carrier density is larger than the temperature rise of the active layer [36]. The changes in the spectral width and side-mode suppression ratio have yet to be determined.

The radiation hardness of laser diodes is discussed in Chapter 6 on the optoelectronic switch matrix.

The relative intensity noise is discussed in Chapter 5 on RF fiber optic links. The harmonic and intermodulation distortion are discussed in Chapter 6 on the optoelectronic switch matrix and also Chapter 5 on RF fiber optic links.

REFERENCES

1. Hecht, J., *The Laser Guidebook*, New York, McGraw-Hill, 1986.
2. Willardson, R.K., and A.C. Beer, (eds.), *Semiconductors and Semimetals*, Vol. 14, New York, Academic Press, 1979, pp. 10, 26, 84.
3. Kawana, A., T. Miyashita, M. Nakahara, M. Kawachi, and T. Hosaka, "Fabrication of Low-Loss Single-Mode Fibers," *Electronics Letters*, Vol. 13, No. 7, March 1977, pp. 188–189.
4. Miya, T., Y. Terumuna, T. Hosaka, and T. Miyashita, "Ultimate Low-Loss Single-Mode Fiber at 1.55 μm," *Electronics Letters*, Vol. 15, No. 4, February 1979, pp. 106–108.
5. Niizeki, N., "Single-Mode Fiber at Zero-Dispersion Wavelength," *Digest Topical Meeting on Integrated Optics and Guided-Wave Optics*, January 16–18, 1978, pp. MB1-1 to MB1-4.
6. Casey, Jr., H.C., and M.B. Panish, *Heterostructure Lasers, Part A, Fundamental Principles*, New York, Academic Press, 1978, pp. 45, 55, 192.

7. Stern, F., "Calculated Spectral Dependence of Gain in Excited GaAs," *J. Applied Physics*, Vol. 47, No. 12, December 1976, pp. 5382–5386.
8. Casey, Jr., H.C., "Room Temperature Threshold-Current Dependence of GaAs-Al$_x$Ga$_{1-x}$As Double-Heterostructure Lasers on x and Active-Layer Thickness," *J. Applied Physics*, Vol. 49, No. 7, July 1978, pp. 3684–3692.
9. Casey, Jr., H.C., M.B. Panish and J.L. Merz, "Beam Divergence of the Emission from Double-Heterostructure Injection Lasers," *J. Applied Physics*, Vol. 44, No. 12, December 1973, pp. 5470–5475.
10. Ikegami, T., "Reflectivity of Mode at Facet and Oscillation Mode in Double-Heterostructure Injection Lasers," *IEEE J. Quantum Electronics*, Vol. QE-8, No. 6, June 1972, pp. 470–476.
11. Butler, J.K., and H. Kressel, "Design Curves for Double Heterojunction Laser Diodes," *RCA Review*, Vol. 38, December 1977, pp. 542–558.
12. Casey, H.C., Jr., and M. B. Panish, *Heterostructure Lasers, Part B, Materials and Operating Characteristics*, New York, Academic Press, 1978, p. 252.
13. Botez, D., "Near and Far-Field Analytical Approximations for the Fundamental Mode in Symmetric Waveguide Double Heterostructure Lasers," *RCA Review*, Vol. 39, December 1978, pp. 577–603.
14. Sze, S.M., *Physics of Semiconductor Devices*, Second Edition, New York, John Wiley and Sons, 1981, p. 716.
15. Murakami, Y., J.I. Yamada, J.I. Sakai, and T. Kimura, "Microlens Tipped on a Single-Mode Fiber End for InGaAsP Laser Coupling Improvement," *Electronics Letters*, Vol. 16, No. 9, April 1980, pp. 321–322.
16. Eisenstein, G., and D. Vitello, "Chemically Etched Conical Microlenses for Coupling Single-Mode Lasers into Single-Mode Fibers," *Applied Optics*, Vol. 21, No. 19, October 1982, pp. 3470–3474.
17. Sumida, M., and K. Takemoto, "Lens Coupling of Laser Diodes to Single-Mode Fibers," *IEEE J. Lightwave Technol.*, Vol. LT-2, No. 3, June 1984, pp. 305–311.
18. Khoe, G.D., and H.G. Kock, "Laser-to-Monomode-Fiber Coupling and Encapsulation in a Modified TO-5 Package," *IEEE Trans. Electron Devices*, Vol. ED-32, No. 12, December 1985, pp. 2707–2712.
19. Yariv, A., *Optical Electronics*, 3rd Edition, New York, Holt, Rinehart and Winston, 1985, p. 490.
20. Tsang, W.T., (ed.), *Semiconductors and Semimetals*, Vol. 22, *Lightwave Communications Technology, Part B, Semiconductor Injection Lasers, I*, Orlando, Academic Press, Inc., 1985, p. 71.
21. Lau, K.Y., N.B. Chaim, and I. Ury, "Wideband Laser Diodes Spark New Designs," *Microwaves & RF*, Vol. 23, No. 11, November 1984, pp. 109–116.
22. Lau, K.Y., N.B. Chaim, I. Ury, C. Harder and A. Yariv, "Direct Amplitude Modulation of Short-Cavity GaAs Lasers upto X-Band Frequencies," *Applied Physics Letters*, Vol. 43, No. 1, July 1983, pp. 1–3.
23. Wakao, K., N. Takagi, K. Shima, K. Hanamitsu, and K.I. Hori, "Catastrophic Degradation Level of Visible and Infrared GaAlAs Lasers," *Applied Physics Letters*, Vol. 41, No. 12, December 1982, pp. 1113–1115.
24. Blauvelt, H., S. Margalit, and A. Yariv, "Large Optical Cavity AlGaAs Buried Heterostructure Window Lasers," *Applied Physics Letters*, Vol. 40, No. 12, June 1982, pp. 1029–1031.
25. Tucker, R.S., and D.J. Pope, "Microwave Circuit Models of Semiconductor Injection Lasers," *IEEE Trans. Microwave Theory Tech.*, Vol. MTT-31, No. 3, March 1983, pp. 289–294.
26. Touchstone, Circuit Analysis and Optimization Software, EEsof, Inc., 1987.
27. Tsukada, T., "GaAs-GA$_{1-x}$Al$_x$As Buried-Heterostructure Injection Lasers," *J. Applied Physics*, Vol. 45, No. 11, November 1974, pp. 4899–4906.
28. SL-Series Single-Mode Lasers, Preliminary Data Sheet, Ortel Corporation.

29. Wakao, K., K. Nakai, T. Sanada, M. Kuno, T. Odagawa, and S. Yamakoshi, "InGaAsP/InP Planar Buried Heterostructure Lasers with Semi-Insulating InP Current Blocking Layers Grown by MOCVD," *IEEE J. Quantum Electronics*, Vol. QE-23, No. 6, June 1987; pp. 943–947.
30. Bowers, J.E., B.R. Hemenway, T.J. Bridges, E.G. Burkhardt, and D.P. Wilt, "26.5 GHz Bandwidth InGaAsP Lasers with Tight Optical Confinement," *Electronics Letters*, Vol.. 21, No. 23, November 1985, pp. 1090–1091.
31. Zah, C.E., J.S. Osinski, S.G. Menocal, N. Tabatabaie, T.P. Lee, A.G. Dentai, and C.A. Burrus, "Wide-Bandwidth and High-Power 1.3 μm InGaAsP Buried Crescent Lasers with Semi-Insulating Fe-Doped InP Current Blocking Layers," *Electronics Letters*, Vol. 23, No. 1, January 1987, pp. 52–53.
32. Kamite, K., H. Sudo, M. Yano, H. Ishikawa, and H. Imai, "Ultra-High Speed InGaAsP/InP DFB Lasers Emitting at 1.3 μm Wavelength," *IEEE J. Quantum Electronics*, Vol. QE-23, No. 6, June 1987, pp. 1054–1058.
33. H. Ishikawa, H. Soda, K. Wakao, K. Kihara, K. Kamite, Y. Kotaki, M. Matsuda, H. Sudo, S. Yamakoshi, S. Isozumi, and H. Imai, "Distributed Feedback Laser Emitting at 1.3 μm for Gigabit Communication Systems," *IEEE J. Lightwave Technol.*, Vol. LT-5, No. 6, June 1987, pp. 848–854.
34. Yano, M., Y. Kotaki, H. Ishikawa, S. Yamakoshi, H. Imai, T. Chikama, and T. Touge, "Extremely Low-Noise Facet-Reflectivity-Controlled InGaAsP Distributed-Feedback Lasers," *IEEE J. Lightwave Technol.*, Vol. LT-4, No. 10, October 1986, pp. 1454–1459.
35. Furuhama, Y., K. Yasukawa, K. Kashiki, and Y. Hirata, "Present Status of Optical ISL Studies in Japan," *Optical Systems for Space Applications*, SPIE, Vol. 810, 1987, pp. 141–149.
36. Fukada, M., "Laser and LED Reliability Update," *IEEE J. Lightwave Technol.*, Vol. LT-6, No. 10, October 1988, pp. 1488–1495.
37. Lo, Y.T., and S.W. Lee (eds.), *Antenna Handbook: Theory, Applications and Devices*, New York, Van Nostrand Reinhold, 1988, pp. 5–14.

Chapter 3
Electro-Optic Modulators

3.1 INTRODUCTION

Wideband, microwave fiber optic communication systems require high-speed integrated-optic devices and single-mode optical fiber technology for advancement. These high-speed integrated-optic devices, such as modulators, must be capable of converting an input microwave signal to a modulated optical signal. Compared with direct semiconductor laser modulation, external modulation suffers additional coupling losses between the fiber and the modulator and also requires higher RF drive power. However, the useful bandwidth of directly modulated links is limited by the presence of the relaxation oscillation of the laser diode. The relaxation oscillation of commercially available laser diodes is a few GHz. In addition, the laser, although operating with a single longitudinal mode under dc excitation, tends to display a dynamic shift of the peak emission wavelength (chirp) when RF modulated. Therefore, for wideband applications, a CW laser in combination with an integrated-optic modulator is preferred. The bandwidth of such a link is expected to extend to several tens of GHz.

This chapter presents the basic principles of an integrated-optic modulator based on a Ti-diffused $LiNbO_3$ Mach-Zehnder interferometer with traveling wave asymmetric coplanar stripline electrodes. First, there is a need to understand electromagnetic wave propagation in electro-optic crystals. Second, understanding the propagation parameters of planar transmission media on dielectric substrates is also important. Therefore, Sections 3.2 and 3.3 present the optical waveguide fundamentals and electro-optic control, respectively. Section 3.4 describes the electrode structure, while Section 3.5 discusses electrode characteristic impedance and attenuation. In a practical situation, optical power is coupled in and out of the modulator through a single-mode fiber; hence, the coupling between a single-mode optical fiber and an optical waveguide is discussed in Section 3.6. Finally, Section 3.7 presents the modulator design considerations and typical performance characteristics.

3.2 OPTICAL WAVEGUIDE FUNDAMENTALS

This section is devoted to the basic theory of thin-film optical waveguides and its relation to the design of modulators. Figure 3.1 presents a schematic of a buried channel waveguide. In this figure, n_f and n_s denote the indices of the guiding layer and substrate, respectively. Light spreading due to diffraction on the guide surface is effectively controlled in buried channel waveguides. Modulators require a buried-channel waveguide in which light is transversely confined in the y-direction in addition to being confined along the depth. The advantage of the buried channel guide is its smooth surface and the easy location of planar electrodes required to achieve light modulation.

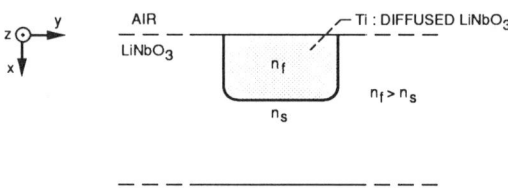

Figure 3.1 Schematic of a typical buried optical waveguide.

3.2.1 Single-Mode Optical Waveguides

Optical channel waveguides surrounded by dissimilar dielectric materials, cannot support pure TE or TM modes, but they support two families of hybrid modes. The hybrid modes are polarized in the x or y direction. The guided modes supported by these guides are therefore classified depending on whether the main electric field component lies in the x or y direction. The mode having the main electric field E_x is called the E_{pq}^x mode [1,2]. The subscripts p and q denote the number of nodes of the electric field E_x in the x and y directions, respectively. This mode resembles the TM mode in a slab waveguide and is sometimes called a *TM-like mode*. Similarly, the E_{pq}^y mode (*TE-like mode*) has the main electric field E_y.

In modulator design, the optical guide must support only the fundamental E_{00}^x or E_{00}^y mode. Ti-diffused LiNbO$_3$ waveguides commonly used in modulators have a Gaussian index profile in both the x and y directions, as shown in Figure 3.2. In this figure, d_x and d_y denote the diffusion depths measured at the intensity $1/e$ points. In general, the refractive indices of the cladding layer, guiding layer, and substrate are denoted as n_c, n_f, and n_s, respectively. However, the cladding layer is absent in the case of modulators discussed in this chapter. By solving the electromagnetic boundary value problem for this type of structure, we can show that for single-mode propagation the ratio of d_x/λ should satisfy the relation [3]:

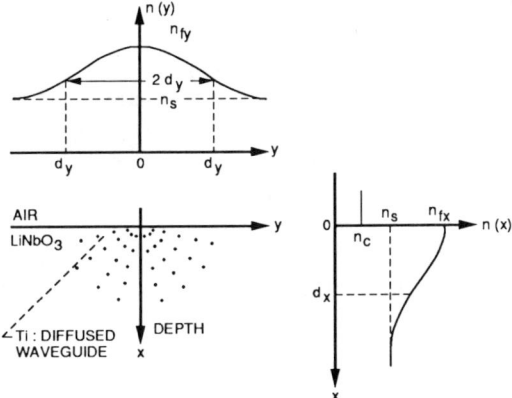

Figure 3.2 Gaussian index profiles in the x and y directions of a graded-index channel waveguide.

$$0.26 < \left(\frac{d_x}{\lambda}\right)(n_s \Delta n)^{0.5} \leq 0.39 \tag{3.1}$$

and

$$2\frac{d_y}{d_x} = 2 \tag{3.2}$$

where Δn is the index change at the guide surface and is equal to $n_f - n_s$. Figure 3.3 illustrates the intensity distribution at any cross section of the buried optical waveguide.

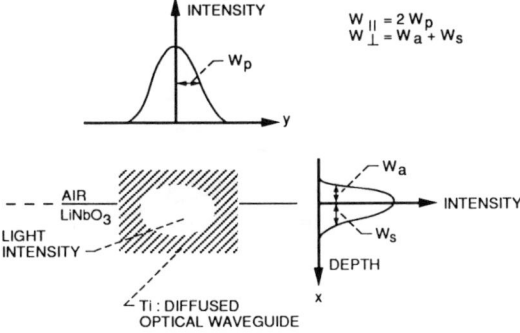

Figure 3.3 Electric field (intensity) distribution in the cross section of the guide.

The process of fabricating the optical waveguides involves first of all sputtering a few hundred Angstrom thick layer of titanium on to the polished surface of a LiNbO$_3$ crystal. Next, the waveguide pattern is photolithographically defined and the crystal is baked at about 1000° C in flowing wet argon gas for four to six hours. During this period, titanium in-diffusion occurs. Next, to compensate for the lack of oxygen, the sample is continued to be heated at 1000° C for one hour in flowing wet oxygen. Finally, the sample is cooled to room temperature in about 15 to 20 minutes.

As a numerical example, consider a Ti-diffused LiNbO$_3$ waveguide operating at 0.63 μm wavelength and the value of n_s and Δn are 2.2 and 0.01, respectively. Then, the value of d_x and d_y should lie in the range of

$$1.1 < d_x(\mu m) < 1.66$$
$$2.2 < 2d_y(\mu m) < 3.31$$

for single-mode propagation.

3.2.2 Single-Mode Optical Branching Waveguide

In single-mode optical branching waveguides, the two quantities of greatest importance are the branching angle ψ and the width of the waveguide W (Figure 3.4). The angle ψ determines the mode separating and recombining functions as well as the scattering losses associated with the fork regions. The angle ψ also determines how long the branching section must be before the coupling between the two arms is negligibly small and, thereby determines the overall length of the device. Because the buried channel waveguide has a diffused (Gaussian) index profile both in the x and y directions an exact analysis is quite cumbersome. Hence we use a simple ray-optics approximation and calculate a critical angle for total reflection of a ray on the diffused sides. The critical angle, θ_{cm} for the mth hybrid mode is given by [4]

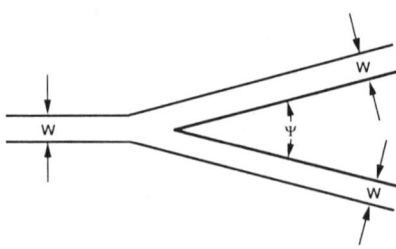

Figure 3.4 Schematic of a single-mode branching waveguide.

$$\sin\theta_{cm} = n_e/\beta_{0m} \qquad (3.3)$$

where

β_{0m} = propagation constant of the mth hybrid mode;
n_e = extraordinary index of LiNbO$_3$.

As a numerical example, if n_e is 2.150 and β_{0m} is $n_e + 0.002$, θ_{cm} is 87.52°, which means that the half-angle of the fork is to be shallower than 2.47° for a wave to follow the guide without suffering appreciable scattering loss. For modulators the distance of separation between the two arms typically ranges from about 5 to 12 μm to keep the coupling small [5].

3.3 ELECTRO-OPTIC CONTROL

3.3.1 Index Ellipsoid

LiNbO$_3$ is a ferroelectric crystal and hence exhibits optical anisotropy. The optical anisotropy is characterized by the symmetric 3 × 3 dielectric tensor. By choosing the x, y, and z axes of an orthogonal coordinate system to coincide with the principal axes of the crystal structure, all off-diagonal elements vanish in the dielectric tensor, and as a result [3,6]:

$$[\varepsilon] = \begin{bmatrix} \varepsilon_{11} & 0 & 0 \\ 0 & \varepsilon_{22} & 0 \\ 0 & 0 & \varepsilon_{33} \end{bmatrix} \qquad (3.4)$$

Thus, the dielectric response of the crystal can be described by the permeability tensor, defined by

$$\mathbf{D} = \varepsilon_0[\varepsilon]\mathbf{E} \qquad (3.5)$$

The stored energy in the crystal is

$$w_e = \frac{1}{2}\mathbf{E} \cdot \mathbf{D} \qquad (3.6)$$

Substituting (3.4) and (3.5) into (3.6) yields

$$w_e = \frac{1}{2\varepsilon_0}\left[\frac{D_x^2}{\varepsilon_{11}} + \frac{D_y^2}{\varepsilon_{22}} + \frac{D_z^2}{\varepsilon_{33}}\right] \qquad (3.7)$$

By replacing $D_i/\sqrt{2\,\varepsilon_0 w_e}$ with i (where i is equal to x, y, z), (3.7) reduces to

$$\frac{x^2}{n_x^2} + \frac{y^2}{n_y^2} + \frac{z^2}{n_z^2} = 1 \tag{3.8}$$

where $\varepsilon_{ij} = n_i^2$. This equation represents the index ellipsoid with the principal axes x, y, and z. The semiaxes of the index ellipsoid along the x, y, and z directions equal the principal refractive indices n_x, n_y, and n_z, respectively. In optically isotropic crystals, $n_x = n_y = n_z$. In optically uniaxial crystals, $n_x = n_y \neq n_z$. The z axis of uniaxial crystals is also known as the optical axis. LiNbO$_3$ is an example of a uniaxial crystal for which $n_x = n_y = n_o$, $n_z = n_e$ and $n_o > n_e$. Here the ordinary and extraordinary indices of refraction are denoted as n_o and n_e, respectively.

3.3.2 Electro-Optic Effect

The refractive index of a crystal changes due to the electro-optic effect when an electric field is applied. The linear electro-optic effect, called the *Pockels effect*, in which the index varies linearly with the applied electric field, is used for light modulation. In the presence of an externally applied electric field, the index ellipsoid is expressed as [3,6]:

$$B_{11}x^2 + B_{22}y^2 + B_{33}z^2 + 2B_{23}yz + 2B_{31}zx + 2B_{12}xy = 1 \tag{3.9}$$

where x, y, and z generally are not the principal axes. If we choose the axes to be parallel to the principal dielectric axes of the crystal with zero applied field, (3.9) reduces to (3.8). The constants B_{ij} of the index ellipsoid are related to the vector $\mathbf{E}^e$ of the applied electric field through the 6 × 3 electro-optic coefficient tensor, as follows [3]:

$$\begin{bmatrix} B_{11} - 1/n_x^2 \\ B_{22} - 1/n_y^2 \\ B_{33} - 1/n_z^2 \\ B_{23} \\ B_{31} \\ B_{12} \end{bmatrix} = \begin{bmatrix} r_{11} & r_{12} & r_{13} \\ r_{21} & r_{22} & r_{23} \\ r_{31} & r_{32} & r_{33} \\ r_{41} & r_{42} & r_{43} \\ r_{51} & r_{52} & r_{53} \\ r_{61} & r_{62} & r_{63} \end{bmatrix} \begin{bmatrix} E_x^e \\ E_y^e \\ E_z^e \end{bmatrix} \tag{3.10}$$

where [r] is an intrinsic characteristic of the material. In LiNbO$_3$, the electro-optic coefficient tensor [r] is

$$[r] = \begin{bmatrix} 0 & -r_{22} & r_{13} \\ 0 & r_{22} & r_{13} \\ 0 & 0 & r_{33} \\ 0 & r_{51} & 0 \\ r_{51} & 0 & 0 \\ -r_{22} & 0 & 0 \end{bmatrix} \qquad (3.11)$$

3.3.3 Index Changes

Consider the case when the electric field is applied to a LiNbO$_3$ crystal along the optical axis (i.e., $E_x^e = E_y^e = 0$ and $E_z^e \neq 0$), and light propagates along the x-axis, as shown in Figure 3.5. From (3.10) and (3.11), we obtain

$$\begin{aligned} B_{11} &= 1/n_x^2 + r_{13}E_z^e \\ B_{22} &= 1/n_y^2 + r_{13}E_z^e \\ B_{33} &= 1/n_z^2 + r_{33}E_z^e \\ B_{23} &= B_{31} = B_{12} = 0 \end{aligned} \qquad (3.12)$$

Substituting (3.12) into (3.9) results in

$$(1/n_x^2 + r_{13}E_z^e)x^2 + (1/n_y^2 + r_{13}E_z^e)y^2 + (1/n_z^2 + r_{33}E_z^e)z^2 = 1 \qquad (3.13)$$

In the $y - z$ plane, for $x = 0$ (Figure 3.5), the above equation reduces to

$$(1/n_o^2 + r_{13}E_z^e)y^2 + (1/n_e^2 + r_{33}E_z^e)z^2 = 1 \qquad (3.14)$$

In a manner similar to (3.8), the new $1/n_y^2$ and $1/n_z^2$ can be read from the above equation as

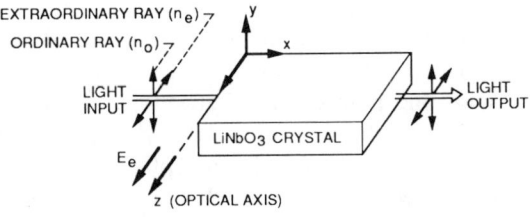

Figure 3.5 Schematic indicating the electric field direction along the optical axis of a Y-cut, x-propagating LiNbO$_3$ crystal.

$$1/n_y^2 = 1/n_o^2 + r_{13}E_z^e \qquad (3.15a)$$

$$1/n_z^2 = 1/n_e^2 + r_{33}E_z^e \qquad (3.15b)$$

Equation (3.15), when solved for n_y and n_z, results in

$$n_y = n_o(1 + n_o^2 r_{13}E_z^e)^{-1/2} \qquad (3.16a)$$

$$n_z = n_e(1 + n_e^2 r_{33}E_z^e))^{-1/2} \qquad (3.16b)$$

By expanding the above equation in a binomial series and taking the first term only, we obtain

$$n_y = n_o(1 - 0.5\, n_o^2 r_{13}E_z^e)$$
$$= n_o - \Delta n_o \qquad (3.17)$$

$$n_z = n_e(1 - 0.5\, n_e^2 r_{33}E_z^e)$$
$$= n_e - \Delta n_e \qquad (3.18)$$

The second term, Δn_o or Δn_e, represents the index changes due to the Pockels effect, determined by the electro-optic coefficients and the strength of the applied electric field. The sign of the index change also depends on the polarity of the voltage applied to the crystal. The largest index change is obtained by applying an electric field along the optical axis of the LiNbO$_3$ crystal, since for that crystal r_{33} is the largest electro-optic coefficient. The maximum value of the index change is 1.6×10^{-3}, limited by the breakdown electric field of LiNbO$_3$, which is about 10 V/μm. Table 3.1 presents the Pockels constant of LiNbO$_3$.

3.4 ELECTRODE STRUCTURE

The largest electro-optic coefficient r_{33} can be utilized for an index change only when the guided mode is polarized along the optical axis. To achieve this condition, two different types of electrode configurations are considered, depending on the cutting of the LiNbO$_3$ crystal as shown in Figure 3.6. In Y-cut LiNbO$_3$, two planar electrodes can be used, with one placed on either side of the optical waveguide. This results in an electric field, E_z^e that is horizontal and also parallel to the optical axis [7]. In Z-cut LiNbO$_3$, the planar electrodes are placed on top of the optical waveguide. This results in an electric field, E_z^e, that is vertical and also parallel to the optical axis [7]. The Z-cut LiNbO$_3$ has the advantage that the E_{pq}^z mode (TM-like), the polarization of which is normal to the guide surface, can propagate without any mode conversion, even if the device involves bends and branches. However, because

Table 3.1
Linear Electro-Optic Coefficients of LiNbO$_3$

Symmetry	Pockels Constants $\times 10^{-12}$ (m/V)[a]	Refractive Index	Wavelength Range for Transparency (μm)	Specific Dielectric Constants[c]	Curie Temperature (°C)	Temperature Coefficient of the Refractive Indices[d] ($\delta n/\delta T$) $\times 10^{-5}$
3m	$r_{33} = 30.8$	$n_o = 2.286$[a]	0.4–5	$\varepsilon_\perp = 43$	1470	n_e: +5.3
	$r_{13} = 8.6$	$n_e = 2.200$[a]		$\varepsilon_\parallel = 28$		n_o: +0.56
	$r_{22} = 3.4$	$n_o = 2.237$[b]				
	$r_{51} = 28$	$n_e = 2.157$[b]				

[a] At $\lambda = 0.6328\ \mu$m, room temperature under constant strain.
[b] At $\lambda = 1.0\ \mu$m.
[c] $\varepsilon_\perp$ and $\varepsilon_\parallel$ are values normal and parallel to the crystal axis, respectively.
[d] Over the temperature range of 20–40° C.

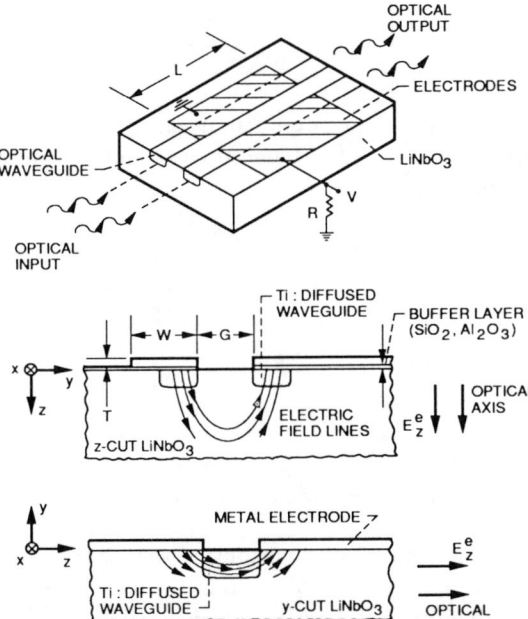

Figure 3.6 Planar electrode structure for LiNbO$_3$ waveguide modulators in which the guided modes are controlled by the electro-optic coefficient r_{33}.

the metal electrodes are placed directly on the optical waveguide, a buffer layer is required to minimize the propagation loss [8].

3.5 ELECTRODE IMPEDANCE AND ATTENUATION

Figure 3.7 illustrates the geometry of asymmetric coplanar stripline electrodes. The characteristic impedance Z_0 of the asymmetric coplanar stripline can be approximately obtained by a conformal mapping technique which makes use of the Schwartz-Christoffel transformation [9]. The simplifying assumptions are made that the thickness T of the metalization is zero and the structure is lossless. This technique leads to the following closed-form expression for the characteristic impedance:

$$Z_0 = \frac{240\pi}{\sqrt{\varepsilon_{\text{eff}}}} \frac{K(t_a)}{K(t_b)} \qquad (3.19)$$

where

$$t_a = \sqrt{G/(2W + G)} \qquad (3.20a)$$

$$t_b = \sqrt{1 - t_a^2} \qquad (3.20b)$$

ε_{eff} = effective dielectric constant

$$= (1 + \sqrt{\varepsilon_\| \varepsilon_\perp})/2 \qquad (3.21)$$

$K(t_a)$ and $K(t_b)$ are the complete elliptic integral of the first kind. The value of $\varepsilon_\|$ and $\varepsilon_\perp$ can be found in Table 3.1. The computed and the measured values of Z_0 are in fair agreement [9].

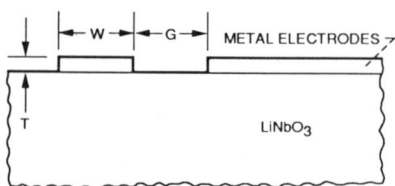

Figure 3.7 An asymmetric coplanar stripline electrode structure.

A more accurate technique of determining the Z_0 is to use an integral equation approach [10]. This technique is also capable of taking into consideration the finite thickness of the metalization and besides yields the attenuation per unit length, α of the structure. The disadvantage of this technique is that it is computationally complex. The computed characteristic impedance and the normalized attenuation of asymmetric coplanar strip line on LiNbO$_3$ as a function of W/G and T/G obtained by the above technique [10] is presented in Figure 3.8. In this figure, R_s is the sheet resistance of the electrode metalization in Ω/square. Further, the curves are strictly valid only at frequencies where the skin depth d is much less than the electrode thickness, T. In this domain, R_s is expressed as [10]:

$$R_s \approx 1/d\sigma = \sqrt{\frac{\pi\mu}{\sigma}} f^{0.5} \tag{3.22a}$$

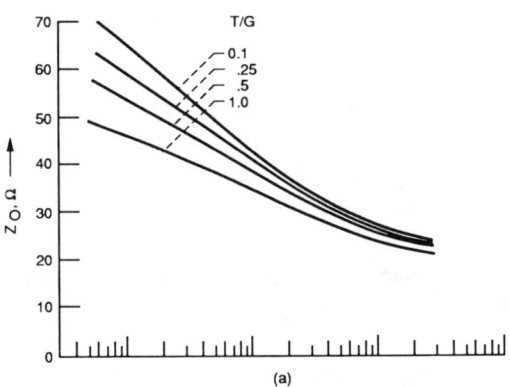

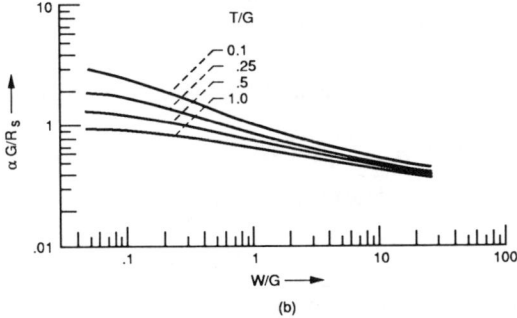

Figure 3.8 (a) Computed characteristic impedance and (b) normalized attenuation of asymmetric coplanar stripline on LiNbO$_3$ as a function of W/G and T/G.
Source: Donnelly, J.P., and A. Gopinath, "A Comparison of Power Requirements of Traveling Wave LiNbO$_3$ Optical Couplers and Interferometric Modulators," *IEEE J. Quantum Electronics*, Vol. QE-23, No. 1, January 1987, pp. 30–41. Reprinted with permission.

where

σ = conductivity;
μ = permeability of the metal.

At low frequencies, where the skin depth is large compared to the electrode thickness, a reasonable approximation for R_s is [10]:

$$R_s = 2/\sigma T \qquad (3.22b)$$

As a numerical example, let us consider an asymmetric coplanar strip line on LiNbO$_3$ with W equal to 20 μm, G equal to 4 μm and T equal to 2 μm. The ratios W/G and T/G are 5 and 0.5, respectively. The value of $\alpha G/R_s$ from Figure 3.8 is approximately 0.07. Assuming the electrode metal to be aluminum, σ is 3.82 × 10^7 mhos/meter, and therefore d is 2.57 $f^{-0.5}$ μm, where f is in GHz. In the low-frequency domain ($f < 1$ GHz), α is 4.58 dB/cm, and at high frequencies ($f > 4$ GHz) α is 1.78 dB/cm $\cdot \sqrt{\text{GHz}} \cdot f^{0.5}$.

3.6 COUPLING LOSS BETWEEN A SINGLE-MODE OPTICAL FIBER AND A CHANNEL WAVEGUIDE

The depth index profile of a conventional Ti-diffused LiNbO$_3$ channel waveguide is asymmetric, and consequently the waveguide depth mode is also quite asymmetric, as illustrated in Figure 3.3. However, single-mode fibers have circular symmetric mode profiles. Thus, the fiber-waveguide coupling efficiency is limited by the mismatch between the fiber and waveguide mode.

The coupling efficiency can be calculated using the equation [11]

$$\eta = \frac{2[(1/a^2 + 1/W_a^2)^{-0.5} + (1/a^2 + 1/W_s^2)^{-0.5}]^2}{a^2 W_p (W_a + W_s)(1/a^2 + 1/W_p^2)} \qquad (3.23)$$

where

a = mode size of the single-mode fiber;
W_a = upper half of the waveguide mode depth;
W_s = lower half of the waveguide mode depth;
W_p = width of the waveguide mode;
r = W_a/W_s = depth mode assymetry ratio;
γ = $W_\parallel/W_\perp$ = is the waveguide mode eccentricity;
$W_\parallel$ = $2 W_p$ = full width at the intensity $1/e$ point;
$W_\perp$ = $W_a + W_s$ = depth at the intensity $1/e$ point

As a numerical example, let us suppose that r and γ are 0.6 and 1.5, respectively. Then, the coupling loss is about 0.4 dB per fiber-waveguide interface. The assumption made in this computation is that the waveguides mean mode size $(W_\| W_\perp)^{0.5}$ is exactly equal to the single-mode fiber size a. If r is increased to 0.8, the coupling loss reduces to about 0.2 dB per fiber-waveguide interface.

3.7 INTERFEROMETRIC WAVEGUIDE MODULATORS

3.7.1 Principle of Operation

Figure 3.9 illustrates a Mach-Zehnder interferometric traveling wave electro-optic modulator. Note that this modulator uses the free space interferometer principle proposed at the end of the nineteenth century [12,13]. In the structure of Figure 3.9, light entering the device is divided at the first Y-junction equally into the two arms. A pair of electrodes are placed on the two parallel arms. By applying a voltage V_0 to the electrodes, the guided mode propagating along the upper arm undergoes a phase shift and interferes with the reference guided mode propagating along the lower arm at the output Y-junction. The output light intensity is thus modulated in response to the phase difference between these two guided modes [14].

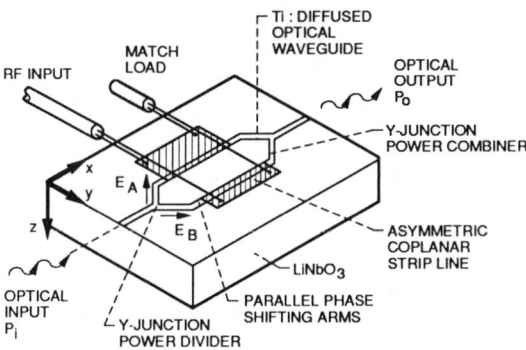

Figure 3.9 Schematic of a traveling wave Mach-Zehnder type of electro-optic modulator.

In the modulator shown in Figure 3.9, when a Z-cut LiNbO$_3$ is used as the substrate and the E_{pq}^z mode is launched, the guided modes are phase shifted by $+\Delta\phi$ and $-\Delta\phi$ in the upper and lower arms, respectively. Thus, the net phase difference between the two guided modes is $2\Delta\phi$. The above technique of obtaining a phase shift is known as *push-pull operation*. The phase shift $2\Delta\phi$ is related to the electrode length L and the distance of separation G as follows:

$$2\Delta\phi = \pi \frac{V_0}{V_\pi} \qquad (3.24)$$

where

$$V_\pi = \frac{\lambda G}{2\Gamma n_e^3 r_{33} L} \qquad (3.25)$$

In the above equation, V_π denotes the half-wave voltage or the voltage required for 180° phase shift. The overlap integral Γ takes into account the index change induced in the waveguide by the amount of overlap between the externally applied electric field and the optical field of the E_{pq}^z mode. The overlap integral calculated [7] for the asymmetric coplanar strip line as a function of the electrode gap G to the optical mode width $W_\parallel$, is shown in Figure 3.10.

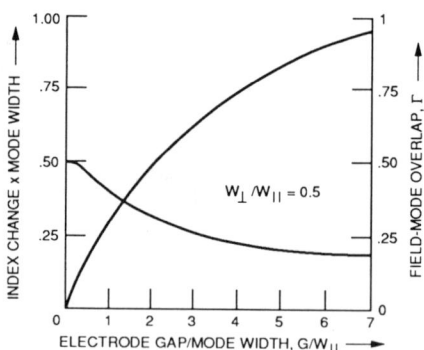

Figure 3.10 Calculated overlap integral *versus* electrode gap divided by optical mode width ratio for asymmetric coplanar stripline.
Source: Alferness, R.C., "Waveguide Electro-optic Modulators," *IEEE Trans. Microwave Theory Tech.*, Vol. MTT-30, No. 8, August 1982, pp. 1221–1137. (Reprinted with permission.)

3.7.2 Extinction Ratio

Let us suppose that the structure of Figure 3.9 is lossless and that the incident optical power P_i is divided between the upper and the lower arms of the input Y-junction with electric field amplitudes E_A and E_B, respectively. That is,

$$|E_A|^2 + |E_B|^2 = P_i \qquad (3.26)$$

Further, the power division ratio, r_p can be written as

$$r_p = \frac{|E_A|^2}{|E_B|^2} \tag{3.27}$$

Since the phase difference between the two arms is $2\Delta\phi$, the optical power at the output port can be written as

$$P_o = \frac{1}{2}(|E_A| - |E_B|)^2 + 2|E_A||E_B|\cos^2\Delta\phi \tag{3.28}$$

Eliminating $|E_A|$ and $|E_B|$ by using equations 3.26 and 3.27 yields

$$P_o = \frac{P_i}{2}\left\{\frac{(1-\sqrt{r_p})^2}{1+r_p} + \frac{4\sqrt{r_p}}{1+r_p}\cos^2[(\pi/2)(V_0/V_\pi)]\right\} \tag{3.29}$$

Equation (3.29) suggests that, when the term $\cos^2[(\pi/2)(V_0/V_\pi)]$ is zero, the ratio of P_o/P_i is

$$\frac{P_o}{P_i} = \frac{(1-\sqrt{r_p})^2}{2(1+r_p)} \tag{3.30}$$

and, when it is unity, the ratio of P_o/P_i is

$$\frac{P_o}{P_i} = \frac{(1+\sqrt{r_p})^2}{2(1+r_p)} \tag{3.31}$$

Thus, the output light intensity varies with the applied voltage V_0 as shown in Figure 3.11. The extinction ratio, E_r, is defined as

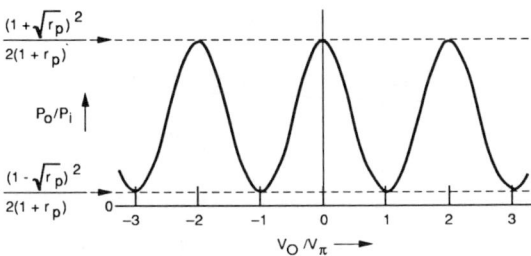

Figure 3.11 Variation of the output light intensity with the applied voltage.

$$E_r = 10 \log \left[\frac{P_o(\max)}{P_o(\min)} \right] \tag{3.32}$$

By utilizing (3.30) and (3.31), (3.32) reduces to

$$E_r = 10 \log \left[\frac{(1 + \sqrt{r_p})}{(1 - \sqrt{r_p})} \right]^2 \text{ dB} \tag{3.33}$$

A good extinction ratio indicates that the chosen Y branch angle, ψ yields nearly ideal 3 dB splitting/combining.

As a numerical example, let us suppose that λ is 0.63 μm, G is 6 μm, and L is 6 mm with Γ approximately equal to 0.3. Then, making use of (3.25) and Table 3.1, V_π is 3.2 V. Further, if r_p is 1.5, E_r from (3.33) is 19.1 dB. This numerical example shows that one of the important advantages of the modulator shown in Figure 3.9 is that the extinction ratio of greater than 10 dB is easily obtained with a low drive voltage, even when the incident power is not equally divided between the two arms due to fabrication errors.

3.7.3 Drive Voltage and Electrode Length Product

A voltage V applied to the electrodes creates an internal field of approximate magnitude, $E \approx V/G$. As a consequence the effective electro-optic induced index change within the optical waveguide from (3.18) is

$$\Delta n = -\frac{n_e^3}{2} r_{33} \frac{V\Gamma}{G} \tag{3.34}$$

In writing this equation, the externally applied electric field is approximated by that of a simple parallel plate capacitor. However, in a practical modulator, the electrodes are coplanar and hence a correction factor is to be applied. This correction factor is given by the overlap integral explained in Section 3.7.1.

The phase shift experienced by the optical signal with the above correction over the interaction length L (electrode length) then is

$$\Delta \phi_0 = -\pi n_e^3 r_{33} \frac{V\Gamma L}{G \lambda}, \quad \text{radians} \tag{3.35}$$

where λ is the free-space optical wavelength. The phase shift required to achieve 100% intensity modulation (i.e., for the power in the two arms to cancel completely at the output) is, in general,

$$\phi_0 = p_\pi \tag{3.36}$$

where p is an integer equal to 1, 3, ... and so forth. The goal in any modulator design is to keep the drive voltage as small as possible for a given length. Hence, equating (3.35) and (3.36) and solving for the voltage length product, we obtain

$$VL = \frac{p\lambda G}{n_e^3 r_{33} \Gamma} \tag{3.37}$$

As a numerical example, for the modulator in the previous example, if p is unity, V is 6.4 V.

3.7.4 Modulator Bandwidth and Electrode Length Product

In deriving (3.37), the assumption is that the light wave travels down the optical waveguide at the same velocity as the microwave drive signal along the metal electrodes. Consequently, the optical wave "sees" the same voltage over the entire electrode length. Thus, we may erroneously conclude that, by choosing the electrode length L to be arbitrarily long, the drive voltage and hence the drive power can be reduced without frequency limitation.

However, in a practical modulator, there is always a small difference between the velocities of the optical wave and the microwave drive signal. This difference in the velocities leads to a reduction or cancellation of the phase shift for arbitrarily large L or drive signal frequency. Considering this difference, the new equation for the phase shift is [7]:

$$\Delta\phi = \Delta\phi_0 \frac{\sin(\theta/2)}{(\theta/2)} \sin[\theta/2 - 2\pi f t_0] \tag{3.38}$$

where

$\theta = \pi f / f_c$
f = microwave signal frequency
$f_c = \dfrac{c}{2 N_m L \delta}$
$c = 3 \times 10^8$ m/s
N_m = refractive index at microwave frequency
  = $\sqrt{\varepsilon_{\text{eff}}}$ = the square root of the effective dielectric constant of an electrical transmission line on LiNbO$_3$),
  = $\left[\dfrac{(1 + \sqrt{\varepsilon_\parallel \varepsilon_\perp})}{2}\right]^{1/2}$
  = 4.2248 for LiNbO$_3$

δ = a measure of the velocity mismatch between the optical and microwave signals,
  = $1 - N_0/N_m$
N_0 = refractive index at optical frequencies taken as 2.2 for $LiNbO_3$ (the extraordinary optical index, n_e).

From (3.38) the frequency for which the phase shift is reduced to 50% of its value for $f = 0$ is [7]:

$$\Delta f L = \frac{2c}{\pi N_m \delta} \qquad (3.39)$$

This equation shows that, for a given electrode length, the achievable bandwidth is critically dependent on the mismatch between the optical and microwave velocities.

As a numerical example, for $LiNbO_3$, the values of N_0 and N_m are 2.2 and 4.2248, respectively. Hence, the product $N_m \delta$ is 2.0248, which results in $\Delta f L$ product of 9.42 GHz-cm. Thus, if we select L as 7.5 mm, Δf is 12.56 GHZ.

3.7.5 Techniques to Reduce Velocity Mismatch

The velocity mismatch between the light wave and microwave signal can be reduced by two techniques. In the first technique, a groove is etched between the parallel optical waveguides. The groove decreases the index of refraction at microwave frequencies and hence improves the match [5]. In the second technique, [15] the electrodes are shifted after an interaction length L_i so that the direction of the applied field in the waveguide is reversed as shown in Figure 3.12. The difference in the time taken by the microwave and optical signals to traverse the entire length, L_i of the electrode is

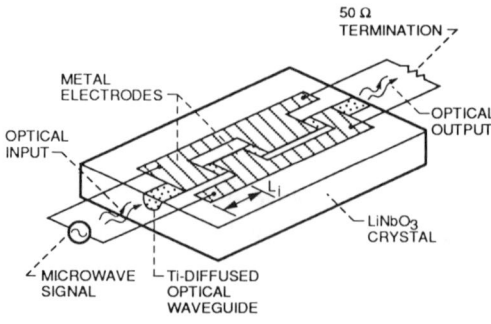

Figure 3.12 Schematic of a traveling wave modulator with periodic phase reversal electrodes.

$$T = \frac{L_i}{c}(N_m - N_0) \qquad (3.40)$$

$$= \frac{L_i}{c}N_m\delta$$

The value of L_i is chosen such that this field direction reversal exactly compensates for the polarity reversal caused by difference between the phase velocities of the optical and microwave signal at a design frequency f_d. This requires that

$$2\pi f_d T = \pi \qquad (3.41)$$

Combining (3.40) and (3.41) results in the effective velocity-match condition [15]:

$$\frac{2\pi N_m f_d L_i \delta}{c} = \pi \qquad (3.42)$$

Note that the technique is not broadband and velocity matching is perfectly achieved only at the design frequency. As a numerical example, for the modulator in the previous example the value of L_i at 12 GHz is 6.20 mm. The induced phaseshift in each section adds in phase when phase reversed electrodes are employed. Consequently, the total interaction length NL_i, where N is the number of electrode section, can be arbitrarily long without the detrimental effect of velocity mismatch. This of course assumes that attenuation is small. Besides, a larger L results in a smaller drive voltage for 100% modulation as per equation 3.37.

3.7.6 Optical Damage and dc Drift

When light in the visible and near-infrared regions of the spectrum is guided in Ti-diffused $LiNbO_3$ waveguide, electrons are excited from impurity levels to the conduction band in the crystal. In the absence of an applied electric field, the optically excited electrons are captured by traps. These traps give rise to positively and negatively charged localized regions, with a spatial electric field, within the crystal. The electric field produces a substantial index change through the linear electro-optic effect of the $LiNbO_3$ crystal. This phenomenon is called optical damage. Optical damage has detrimental effects on bulk, planar-waveguide, and channel-waveguide $LiNbO_3$ devices. An interesting observation made during the course of the above investigations is that thermal annealing the device at 150°C virtually eliminates the effect of optical damage [16].

In the presence of an applied dc electric field, the light-excited electrons move toward the anode. Consequently, there is a reduction in the electric field strength inside the crystal. As a result, the output light intensity from a waveguide device decreases or increases with time. This phenomenon is known as *dc drift*. The time constant is typically a few seconds to a few minutes. In addition, the dc drift is also attributed to the imperfect SiO_2 buffer layer which separates the Ti-$LiNbO_3$ waveguide from the metal electrodes [17]. Improved stability against dc drift has been demonstrated by replacing the SiO_2 buffer layer with an indium tin oxide (ITO) transparent conductor. The long term drift with this change is less than 0.3% in 8 hours [18] for a typical 1.3 μm wavelength modulator.

3.7.7 Radiation Hardness

Ti-diffused $LiNbO_3$ optical waveguide devices were exposed to approximately 10^5 rad of 2.25 MeV β^- particles and 35 keV x-rays. This is a typical radiation environment encountered by satellites in space. At a wavelength of 1.3 μm, no change was detected in transmission or coupling of optical signals between adjacent waveguides either during or after irradiation [19]. This indicates that the refractive index of $LiNbO_3$ is largely unaffected by ionizing radiation.

3.7.8 Experimental Modulator Performance and Discussions

The design and performance of Mach-Zehnder electro-optic traveling wave modulators have been reported by several investigators [5,7,16,20]. These modulators are fabricated on Z-cut $LiNbO_3$ crystal. Typical modulator design parameters and electrical characteristics are summarized in Table 3.2. The computed 3 dB bandwidth using (3.39) is 15.7 GHz, whereas the measured bandwidth is 17 GHz. This discrepancy may be caused by the lowering of the effective dielectric constant due to the presence of a buffer layer. The figure of merit of a modulator is defined as the RF drive power per unit bandwidth. For the modulator under consideration this is 7.06 μW/MHz for 100% intensity modulation [16].

The most serious disadvantage of this device is the limited optical power-handling capability at wavelength near or about 0.63 μm and 0.83 μm. As has been observed, if the optical power to the modulator is increased beyond a certain threshold, typically about 50 μW into a 4 μm wide channel waveguide [15], the $LiNbO_3$ crystal is susceptible to optical damage. The optical damage threshold of $LiNbO_3$ is at least an order of magnitude larger at 1.3 μm than at 0.83 μm [21]. Thus, fiber optic links at 1.3 μm incorporating $LiNbO_3$ modulators and high-speed InGaAs photodiodes may have a substantial performance advantage over links at 0.83 μm.

Table 3.2
Mach-Zehnder Traveling Wave Electro-Optic Modulator Design Parameters and Electrical Characteristics

Wavelength (λ)	0.6328 μm
Ti-Diffused Optical Waveguide Width	4 μm
Substrate	Z-cut LiNbO$_3$
Buffer Layer	SiO$_2$
Optical Waveguide Branch Angle (ψ)	1°
Electrode Metalization	Gold, 3 μm thick
Electrode Separation (G)	16 μm
Electrode Length (parallel arms, L)	6 mm
Asymmetric Coplanar Stripline Width (W)	11 μm
Characteristic Impedance (Z_0)	50 Ω
Drive Voltage for Constant On-Off Modulation	±3.5 V
Drive Power	120 mW
3 dB Bandwidth	17 GHz
Extinction Ratio (E_r)	15 dB
Figure of Merit (Drive Power/Bandwidth)	7.06 μW/MHz

REFERENCES

1. Knox, R.M., "Dielectric Waveguide Microwave Integrated Circuits—An Overview," *IEEE Trans. Microwave Theory Tech.*, Vol. MTT-24, No. 11, November 1976, pp. 806–814.
2. McLevige, W.V., T. Itoh, and R. Mittra, "New Waveguide Structures for Millimeter-Wave and Optical Integrated Circuits," *IEEE Trans. Microwave Theory Tech.*, Vol. MTT-23, No. 10, October 1975, pp. 788–794.
3. Nishihara, H., M. Haruna, and T. Suhara, *Optical Integrated Circuits*, McGraw-Hill, New York, 1989, pp. 35, 99.
4. Ranganath, T.R., and S. Wang, "Ti-Diffused LiNbO$_3$ Branched-Waveguide Modulators: Performance and Design," *IEEE J. Quantum Electronics*, Vol. QE-13, No. 4, April 1977, pp. 290–295.
5. Haga, H., M. Izutsu, and T. Sueta, "LiNbO$_3$ Traveling-Wave Light Modulator/Switch with an Etched Groove," *IEEE J. Quantum Electronics*, Vol. QE-22, No. 6, June 1986, pp. 902–906.
6. Yariv, A., *Optical Electronics*, 3rd Ed., New York, Holt, Rinehart and Winston, 1985, pp. 10, 275.
7. Alferness, R.C., "Waveguide Electro-optic Modulators," *IEEE Trans. Microwave Theory Tech.*, Vol. MTT-30, No. 8, August 1982, pp. 1221–1137.
8. Masuda, M., and J. Koyama, "Effects of a Buffer Layer on TM Modes in a Metal-Clad Optical Waveguide Using Ti-Diffused LiNbO$_3$ c-plate," *Applied Optics*, Vol. 16, No. 11, November 1977, pp. 2294–3000.
9. Kubota, K., J. Noda, and O. Mikami, "Traveling Wave Optical Modulator Using a Directional Coupler LiNbO$_3$ Waveguide," *IEEE J. Quantum Electronics*, Vol. QE-16, No. 7, July 1980, pp. 754–760.

10. Donnelly, J.P., and A. Gopinath, "A Comparison of Power Requirements of Traveling Wave LiNbO$_3$ Optical Couplers and Interferometric Modulators," *IEEE J. Quantum Electronics*, Vol. QE-23, No. 1, January 1987, pp. 30–41.
11. Komatsu, K., S. Yamazaki, M. Kondo, and Y. Ohta, "Low-Loss Broad-Band LiNbO$_3$ Guided-Wave Phase Modulators Using Titanium/Magnesium Double Diffusion Method," *IEEE J. Lightwave Technol.*, Vol. LT-5, No. 9, September 1987, pp. 1239–1245.
12. Zehnder, L., *"Ein neuer Interferenzrefractor," Zeitschr. f. Instrkde.*, Vol. 11, 1981, pp. 275–285.
13. Mach, L., *"Uber einer Interferenzrefractor," Zeitschr. f. Instrkde.*, Vol. 12, 1982, pp. 89–93.
14. Hunsperger, R.G., *Integrated Optics: Theory and Technology*, Springer-Verlag, Berlin, 1982, p. 135.
15. Alferness, R.C., S.K. Korotky, and E.A.J. Marcatili, "Velocity-Matching Techniques for Integrated Optic Traveling Wave Switch/Modulators," *IEEE J. Quantum Electronics*, Vol. QE-20, No. 3, March 1984, pp. 301–309.
16. Gee, C.M., and G.D. Thurmond, "Wideband Traveling-Wave Electro-Optic Modulator," *Optical Technology for Microwave Applications*, SPIE, Vol. 477, 1984, pp. 17–22.
17. Yamada, S., and M. Minakata, "DC drift Phenomena in LiNbO$_3$ Optical Waveguide Devices," *Japan J. Applied Physics*, Vol. 20, No. 4, April 1981, pp. 733–737.
18. Gee, C.M., G.D. Thurmond, H. Blauvelt, and H.W. Yen, "Minimizing dc Drift in LiNbO$_3$ Waveguide Devices," *Applied Physics Letters*, Vol. 47, No. 3, August 1985, pp. 211–213.
19. Drummond, E.I., "Resistance of Ti-LiNbO$_3$ Devices to Ionizing Radiation," *Electronics Letters*, Vol. 23, No. 23, November 1987, pp. 1214–1215.
20. Stephens, W.E., and T.R. Joseph, "System Characteristics of Direct Modulated and Externally Modulated RF Fiber-Optic Links," *IEEE J. Lightwave Technol.*, Vol. LT-5, No. 3, March 1987, pp. 380–387.
21. Blauvelt, H., and H. Yen, "Transmitter and Receiver Design for Microwave Fiber Optic Links," *Optical Technology for Microwave Applications*, SPIE, Vol. 477, 1984, pp. 44–51.

Chapter 4
Photodetectors

4.1 INTRODUCTION

In a fiber optic communication system, a photodetector is used at the receiver to convert the variation of the received optical power into corresponding varying electric current. Because the received optical power at the receiver is usually very weak, the photodetector must meet very high performance requirements. The foremost of these requirements is the responsivity to the emission wavelength of the laser diode being used. In addition, the photodetector must add a minimum of noise to the system, possess sufficiently wide bandwidth, be insensitive to ambient changes, and require modest power.

Photodetectors can be categorized based on their physical structure into junction type and bulk type. The PIN photodiode, the *Schottky-barrier photodiode* and the *avalanche photodiode* (APD) are of the junction type, while the photoconductive detectors are of the bulk type. All of these devices can be fabricated from III-V compound semiconductors.

In the short wavelength region between 0.7 and 0.9 μm, GaAs Schottky-barrier photodiodes and GaAs photoconductive detectors are observed to have superior performance. However, long distance application of this device is limited because of the excessive attenuation of optical fibers at these wavelengths. Consequently, they are employed only in local area networks.

Optical fiber attenuation and dispersion can be minimized by using wavelengths between 1.0 and 1.7 μm. At these wavelengths, PIN and avalanche photodiodes fabricated from ternary and quarternary alloy semiconductor materials, such as InGaAs and InGaAsP perform very well. Also, the highest speed optical modulators [1] and laser diodes [2] have been demonstrated in this wavelength region. Therefore, very long-haul, high-capacity systems operate at long wavelengths.

4.2 PIN PHOTODIODE PRINCIPLE OF OPERATION

A PIN photodiode consists of a *p*-type and *n*-type semiconductor regions separated by an intrinsic layer (I). In normal operation, a sufficiently large reverse-bias voltage is applied across the device so that the intrinsic region is fully depleted of carriers. This reverse bias sets up an electric field equal to the saturation field so that the carriers travel at the saturation velocity during most of the transit (Figure 4.1). The saturation electric field for several semiconductors is presented in Table 4.1.

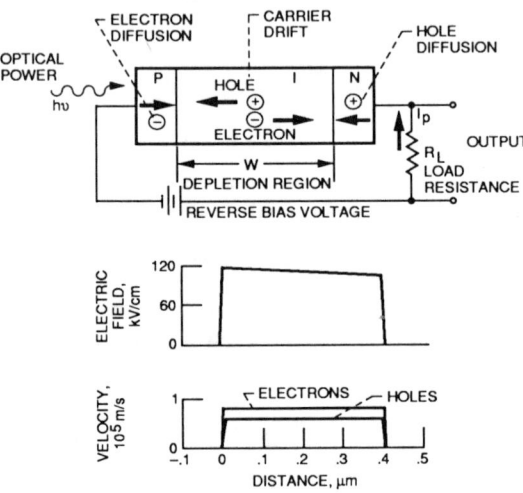

Figure 4.1 Schematic representation of a reverse-biased one dimensional PIN photodiode.

Table 4.1
Material Parameters of Semiconductors at 300K

Material	Mobility μ_o cm²/V-s	Peak Velocity 10^7 cm/s	Relative Dielectric Constant	Band Gap E_g (eV)	Saturation Electrical Field (kV/cm)
GaAs	4500	1.86	12.9	1.439	2.96
InP	3815	2.60	12.30	1.340	4.82
$Ga_{.47}In_{.53}As$	8875	2.2	13.73	0.717	1.61
$Ga_{.27}In_{.73}P_{.4}As_{.6}$	7041	2.7	13.2	0.889	2.51

($N_D = 10^{17}$ donors/cm³)

When a photon of energy equal to or greater than the semiconductor band gap is incident, the photon gives up its energy and excites an electron from the valence band to the conduction band. This process generates free electron-hole pairs, which are known as *photocarriers* (Figure 4.2). The photodiodes are normally designed so that the photocarriers are generated mainly in the depletion region where most of the incident light is absorbed. The high electric field present in the depletion region causes the photocarriers to separate and be collected across the reverse-biased junction. As the carriers traverse the depletion region, a displacement current is induced with one electron flowing for every carrier pair generated. This current flow is known as the *photocurrent*.

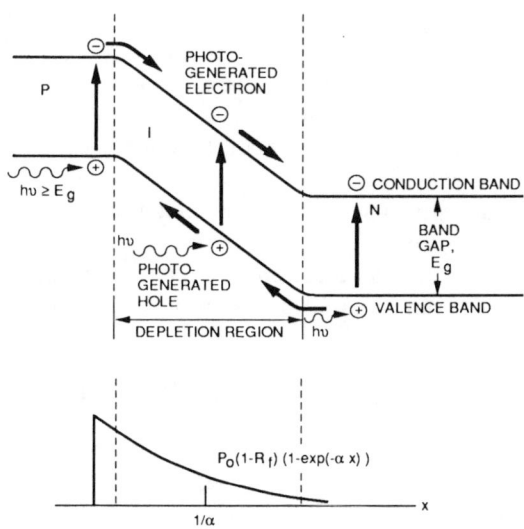

Figure 4.2 Simple energy band diagram for a PIN photodiode.

4.2.1 Quantum Efficiency and Photoresponsivity

As the charge carriers flow through the material, some electron-hole pairs will recombine and hence disappear. On the average, the charge carriers move a distance L_n and L_p for electrons and holes, respectively. This distance is known as the *diffusion length*. The time taken for an electron or hole to recombine is known as the *carrier lifetime*, and is represented by τ_n and τ_p, respectively. The lifetimes and diffusion lengths are related by the expression [3]:

$$L_n = (D_n \tau_n)^{1/2} \tag{4.1}$$

$$L_p = (D_p \tau_p)^{1/2} \tag{4.2}$$

where D_n and D_p are the electron and hole diffusion coefficients, which are expressed in cm²/s.

In a semiconductor material, the optical radiation is absorbed according to the exponential law [3,4]

$$P(x) = P_0[1 - \exp(-\alpha X)](1 - R_f) \tag{4.3}$$

where

α = absorption coefficient per cm at a wavelength λ,
P_0 = incident optical power level,
$P(x)$ = optical power absorbed in a distance x,
R_f = Fresnel (optical) reflection coefficient.

The dependence of the optical absorption coefficient on wavelength is shown in Figure 4.3 for photodiode materials, such as Si, GaAs, InP, InGaAs, and InGaAsP. These curves show that a given semiconductor material is usable only over a limited wavelength range. The upper cutoff occurs when the incident photon is no longer sufficiently energetic to create an electron-hole pair. The upper cut-off wavelength, λ_c, is determined for a given material by the following relation:

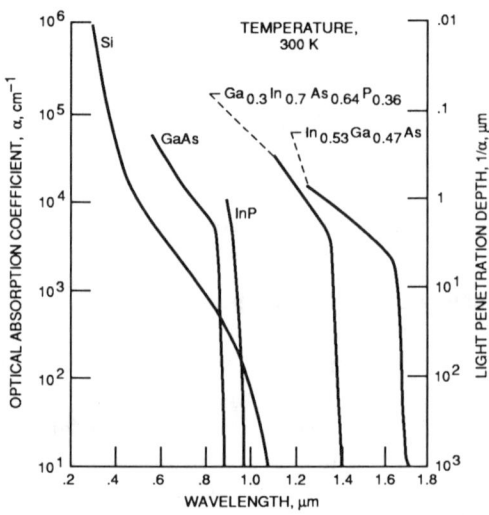

Figure 4.3 Optical absorption coefficient as a function of wavelength for Silicon, Gallium Arsenide, Indium Phosphide, Indium Gallium Arsenide, and Indium Gallium Arsenide Phosphide at 300K.

$$\lambda_c \, (\mu m) = \frac{hc}{E_g} = \frac{1.24}{E_g} \quad (4.4)$$

where

E_g = bandgap of the material in electron volts (eV),
h = Planck's constant,
c = velocity of light in free space.

As a numerical example, the upper cut-off wavelengths of GaAs, InP, and InGaAs at room temperature are 0.87, 0.92, and 1.73 μm, respectively. The band gap of these semiconductors is presented in Table 4.1. The lower cut-off wavelength occurs because the value of α at the shorter wavelength is very large, typically greater than 10^5/cm, and the optical radiation is absorbed very near the surface of the semiconductor where the recombination time is very short. The generated carriers thus recombine before they can traverse the depletion region.

The total optical power absorbed in the depletion region of width W is [3]:

$$P(W) = P_0[1 - \exp(-\alpha W)](1 - R_f) \quad (4.5)$$

The light that enters the photodiode through the p-type layer generates electron-hole pairs within the depletion region or a diffusion length of it. The carriers are separated by the reverse-bias voltage-induced electric field and drift across the depletion region, causing a current flow in the external circuit. Under steady-state conditions, the total current density J_{tot} flowing through the reverse-biased depletion layer is

$$J_{tot} = J_{drift} + J_{diffusion} \quad (4.6)$$

The drift current density is [3]:

$$J_{dr} = \frac{qP_0}{h \nu A}[1 - \exp(-\alpha W)](1 - R_f) \quad (4.7)$$

where

A = photodiode active area,
q = electronic charge,
ν = light frequency.

The surface p-type layer of a PIN photodiode is normally very thin. Thus, diffusion current is principally due to the diffusion of holes from the bulk n-type region. This process can be modeled by a one-dimensional diffusion equation. The solution of this equation gives the diffusion-current density [3]:

$$J_{\text{diff}} = \frac{qP_0(1-R_f)}{Ah\nu}\alpha L_p \frac{\exp(-\alpha W)}{(1+\alpha L_p)} + qP_{n0}\frac{D_p}{L_p} \quad (4.8)$$

where p_{n0} is the thermal equilibrium hole density per cm^3. Because p_{n0} is small, the contribution from the second term is very small and hence, neglected [3]. Thus, the total photogenerated current density through the reverse-biased depletion layer is

$$J_{\text{tot}} = \frac{qP_0(1-R_f)}{Ah\nu}\left[1 - \frac{\exp(1-\alpha W)}{(1+\alpha L_p)}\right] \quad (4.9)$$

The internal quantum efficiency η is the number of electron-hole pairs generated per incident photon of energy $h\nu$ and is given by [3]:

$$\eta = \frac{J_{\text{tot}}A/q}{P_0/h\nu} \quad (4.10)$$

Substituting from (4.9) yields

$$\eta = (1-R_f)\left[1 - \frac{\exp(-\alpha W)}{(1+\alpha L_p)}\right] \quad (4.11)$$

For a high-speed PIN photodiode the product of α and W is far less than 1. Furthermore, the product of α and L_p is also far less than 1. With this approximation the internal quantum efficiency is

$$\eta = (1-R_f)\alpha W \quad (4.12)$$

The transit time cut-off frequency f_t is given by [5]:

$$f_t = 0.553\, v/W \quad (4.13)$$

Where for simplicity the electron and the hole saturation velocities v_n and v_p, respectively, are assumed to be equal and are denoted as v. The saturated carrier velocities for several semiconductor materials are presented in Table 4.1. The transit time bandwidth-efficiency product is given by

$$f_t\eta = 0.553\,\alpha v(1-R_f) \quad (4.14)$$

As a numerical example, if $\alpha = 10^4$/cm at 1.3 μm wavelength for InGaAs, $v = 10^7$ cm/s, and $R_f = 0.1$, the product $f_t\,\eta = 88.5$ GHz. This product is independent of the design parameters. The performance of a photodiode is often characterized by the responsivity, R. The responsivity is a measure of the photocurrent that is generated when optically illuminated and is expressed in A/W:

$$R = \frac{J_{tot}A}{P_0} \quad (4.15a)$$

Substituting from (4.10) yields

$$R = \frac{\eta q}{h\nu} \quad (4.15b)$$

As a numerical example, if the η is equal to 60% at 1.5 μm wavelength, the responsivity is equal to 0.725 A/W.

The quantum efficiency is not a constant at all wavelengths because it varies according to the photon energy $h\nu$. Consequently, the responsivity is a function of the wavelength and photodiode material.

4.2.2 Response Speed

The frequency response of PIN photodiodes is limited primarily by the time required for the photoinjected carriers to drift across the depleted layer and the inherent capacitance of the structure [6]. Hence, for high-speed operation, the depletion layer should be thin so that the transit time is short, and the diode area should be small for small capacitance. However, a very thin depletion layer will decrease the absorption of radiation, and thus diminish the quantum efficiency. A small device area makes difficult fabrication as well as light injection. Thus, the design of a very high speed photodiode involves compromise. Some of the possible trade-offs [6] are indicated in Figure 4.4, which shows 3 dB bandwidth contours in the detector quantum

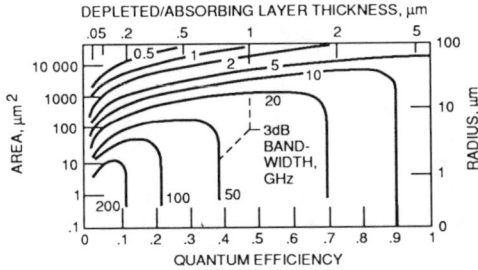

Figure 4.4 Theoretical 3 db bandwidth contours for InGaAs/InP PIN photodiodes (from [6], p. 117. Reprinted with permission).

efficiency-area plane. The computations in Figure 4.4 assume that the absorption coefficient $\alpha = 0.66/\mu m$, the wavelength $\lambda = 1.55$ μm, the sum of the load resistance R_L and detector series resistance R_S are 50 Ω, and the electron and hole saturation velocities, v_n and v_p, are equal to 7×10^6 cm/s and 6.5×10^6 cm/s, respectively [6]. For other load resistances, the vertical axis should be scaled by a factor $50/(R_L + R_S)$. The upper horizontal axis is independent of the absorption coefficient. The lower horizontal axis can be scaled for other wavelengths by using (4.12).

4.2.3 Practical PIN Photodiode Structure

In this section, we describe PIN photodiodes for applications in the long wavelength region. One material suitable for long wavelength operation is the ternary compound $In_xGa_{1-x}As$ (hereinafter denoted as InGaAs). The absorption coefficient curve for this material is shown in Figure 4.3. By adjusting the mole fraction, x, we can achieve a nearly perfect lattice match with the substrate, InP, and simultaneously to adjust the optical responsivity or band gap. A mole fraction of 0.53 yields a lattice which is equal to that of InP and a bandgap of about 0.72 eV (Figure 4.5). For defect-free, thermodynamically stable epitaxial growth of a semiconductor on a semiconductor, substrate lattice-matching is required. If this condition is not met, the epitaxial layer is strained. For thin epitaxial layers, the strained layer has the lattice constant of the substrate and it is called *pseudomorphic*. There is a critical thickness [11] for this layer as a function of lattice mismatch. A pseudomorphic layer thicker than the critical thickness is thermodynamically unstable. That is, the layer can develop defects, either spontaneously, after a heating cycle, or over time.

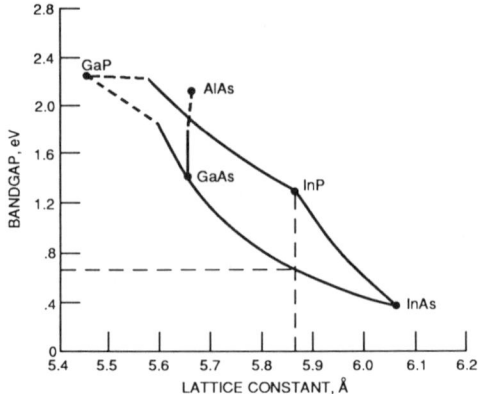

Figure 4.5 Variation of the bandgap as a function of lattice constant for III-V binary and alloy semiconductors.

One advantage of using an InP substrate is that it is transparent to optical illumination in the range of 1.1 to 1.7 μm. A second advantage of InP arises from the ability to monolithically integrate a photodiode and a postdetection MESFET amplifier on a semi-insulating substrate. A semi-insulating substrate also helps in providing higher isolation between devices, and also lower dielectric loss at microwave frequencies.

To achieve high-speed and high quantum efficiency, a heterostructure is used to confine the absorption of light directly in the depletion region [5]. Figure 4.6(a) illustrated a mesa-structure InGaAs-InP PIN photodiode. The structure consists of an n^+ InP substrate which provides a low series resistance to the contacts, an n-type InP buffer layer or window layer (3 μm, $N_D = 5 \times 10^{16}/\text{cm}^3$), an undoped n^- InGaAs absorption layer to absorb the light at 1.3 μm wavelength (1.2 μm, $N_D = 3 \times 10^{14}/\text{cm}^3$), and finally a p^+ InGaAs contact layer. (A Cd-diffused p-n junction is placed at a depth of 0.5 μm below the InGaAs surface.) The diameter of the mesa is about 40 μm. An aperture in the substrate metalization is used to illuminate the

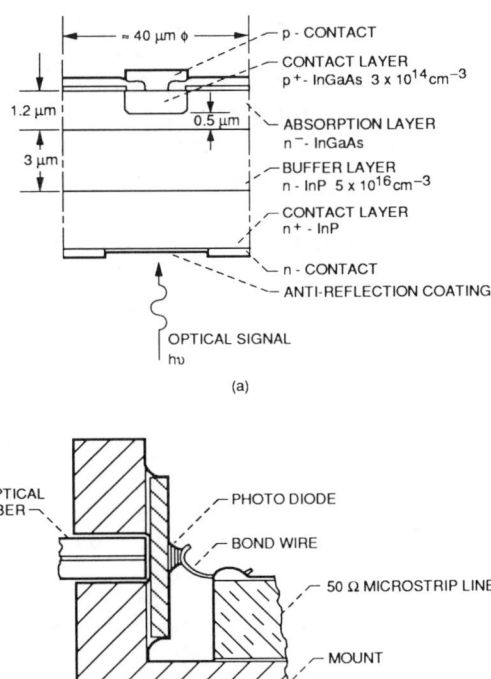

Figure 4.6 (a) Cross-section of InGaAs/InP PIN photodiode; (b) microstrip mount for substrate-side illumination (after [5], p. 117).

diode. A silicon nitride antireflection coating can be applied over the aperture to minimize reflections. Illuminating the photodiode from the backside instead of the topside results in the following advantages [5,7]: first, as the InP layer (substrate) on the backside is transparent between 1.1 and 1.7 μm wavelength, absorption loss and that due to surface recombination are minimized. Second, the light passes twice through the InGaAs absorption region, owing to reflection from the p^+ contact metalization. This permits the use of a thinner InGaAs layer, which reduces transit time without incurring a substantial reduction in internal quantum efficiency. Third, this eliminates the need for a transparent electrode or offset wire-bond pads, as in conventional top-illuminated design, which can introduce bandwidth narrowing parasitic resistance and capacitance respectively. Figure 4.6(b) illustrates a typical coupling arrangement between the optical fiber, photodiode, and microwave circuit.

4.2.4 Leakage Current

The receiver noise is directly proportional to the leakage current. Hence, the leakage current must be kept small. A practical technique to achieve this is to keep the diameter of the mesa small, typically a few tens of μm [5,7]. The resulting small area not only keeps the diode capacitance low, but also minimizes the number of active area defects, which are responsible for the surface leakage current. A typical value of the leakage current is about 5 nA at about -10 V bias [7]. At higher voltages, the leakage current increases exponentially with voltage. This is due to tunneling, which is more pronounced in materials, such as InGaAs, with small band gap and small effective masses. One of the techniques for minimizing tunneling current, even at reasonably high voltages, is to maintain the impurity concentration in the ternary layer below 2×10^{15} cm^{-3} [8]. An alternate technique is to use a very thin layer of a wide-band-gap material, such as InGaAsP, between the absorbing and buffer layers. This technique is explained later in Section 4.5.1 on avalanche photodiodes.

4.2.5 Lumped Element Equivalent Circuit Model and 3 dB Bandwidth

The equivalent circuit model [5] of the mounted photodiode is shown in Figure 4.7. The diode junction capacitance is represented as C_d, and the capacitance due to fring-

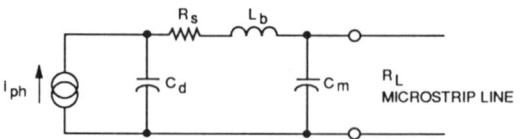

Figure 4.7 Lumped element equivalent circuit of mounted PIN photodiode.

ing fields at the end of the microstrip line is denoted as C_m. The inductance L_b is the inductance of the bond wire. The series resistance R_S is primarily due to the bulk resistance of the low-doped InP buffer layer. The current source of magnitude i_{ph} in parallel with C_d represents the rms photogenerated signal current.

The element values are determined by superposition of the measured reflection coefficient (S_{11}, magnitude and angle) and the reflection coefficient predicted by the above model using circuit analysis and optimization software [9]. As an example, for the PIN photodiode described in Section 4.2.3 and illustrated in Figure 4.6(a), the element values are as follows:

$$C_d = 145 \text{ fF}$$
$$C_m = 15.3 \text{ fF}$$
$$L_b = 159 \text{ pH}$$
$$R_S = 12 \text{ }\Omega$$

The cut-off frequency f_c as predicted by this model is 22 GHz. The transit time cut-off frequency f_t as determined from equation (4.13) is 53.3 GHz. In this computation, the W and v values are taken to be 0.83 μm and 8×10^6 cm/s, respectively. The upper cut-off is the frequency at which the amplitude response of the photodiode drops by 3 dB of its low-frequency value. Thus, the 3 dB bandwidth as determined by the upper cut-off frequency f_0 is a function of both f_c and f_t, and is predicted by the relation [5]:

$$f_0 = \frac{f_c f_t}{(f_c^2 + f_t^2)^{1/2}} \tag{4.16}$$

The predicted value of f_0 is 20.3 GHz. This value is close to the measured value of 17 GHz. InGaAs-InP PIN photodiodes with 3 dB bandwidth as high as 70 GHz have been demonstrated [10].

4.2.6 Spectral Response

The experimental PIN photodiode mounted in the test fixture shown in Figure 4.6(b) did not have an antireflection coating; hence, some amount of the optical power would be reflected back to the source. The reflectivity for InP over the wavelength range of 1.0 to 2.0 μm is about 0.35 [12]. Hence, photodiode responsivity can be estimated by using (4.14) and (4.15). The estimated responsivity is about 0.565 A/W. In doing so the values assumed for α, v, f_t, and λ are 10^4/cm, 8×10^6 cm/s, 53.3 GHz and 1.3 μm, respectively. The measured responsivity of the photodiode at 1.3 μm wavelength is about 0.55 A/W, which is in good agreement with the predicted value.

Finally, Figure 4.8 illustrates as an example a typical measured responsivity and quantum efficiency of a back-illuminated InGaAs-InP PIN photodiode. Notice that the responsivity is almost flat over the entire wavelength range extending from 0.95 to 1.55 μm.

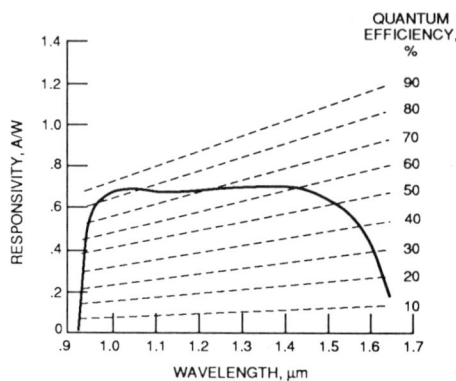

Figure 4.8 Measured response and quantum efficiency of back-illuminated InGaAs/InP PIN photodiode without antireflection coating.
Source: Lee, T.P., C.A. Burrus, A.G. Dentai, and K. Ogawa, "Small area InGaAs/InP PIN photodiodes: fabrication, characteristics and performance of devices in 274 Mb/s and 45 Mb/s lightwave receivers at 1.31 μm wavelength," *Electronics Letters*, Vol. 16, No. 4, February 1980, pp. 155–156.

4.2.7 Signal-to-Noise Ratio and Noise Equivalent Power

Consider an intensity-modulated optical signal given by

$$P(\omega) = P_0 (1 + m\, e^{j\omega t}) \tag{4.17}$$

where

m = modulation index,
ω = modulation frequency,
P_0 = average optical signal power.

The average photocurrent due to the optical signal is

$$I_p = \frac{q\eta P_0}{h\nu} \tag{4.18}$$

For the modulated optical signal, the rms signal power is $mP_0/\sqrt{2}$, and the rms signal current is

$$i_{ph} = q\eta m P_0 / \sqrt{2}\, h\nu \tag{4.19}$$

The principal noise sources associated with the PIN photodiode are the quantum noise or shot noise, the dark current noise, and the surface leakage current noise. The quantum noise or shot noise is due to the statistical nature of the generation and collection of the photoelectrons. As the fluctuations in the number of photocarriers is a fundamental property of the detection process, they set the lower limit on receiver sensitivity when all other conditions are optimized [3]. The quantum noise or shot noise current has a mean square value in a bandwidth B that is proportional to the average value of the photocurrent I_p:

$$\langle i_q^2 \rangle = 2qI_p B \tag{4.20}$$

The photodiode dark current is that which continues to flow through the bias circuit in the absence of illumination. This is a combination of the bulk and surface currents [3]. The bulk dark current i_{db} arises from electrons or holes that are thermally generated at the p-n junction of the photodiode. The mean square value of this noise current is given by

$$\langle i_{db}^2 \rangle = 2qI_d B \tag{4.21}$$

where I_d is the average value of the bulk dark current. The surface dark current i_{ds} is also called a surface leakage current or simply the leakage current. The leakage current is dependent on surface defects, surface area, and bias voltage. The mean square value of this noise current is given by

$$\langle i_{ds}^2 \rangle = 2qI_l B \tag{4.22}$$

where I_l is the average value of the leakage current. In a practical photodiode, the surface leakage current is shunted to ground by a guard ring. The photodiode equivalent resistance R_{eq} contributes a mean square thermal or Johnson noise current, given by

$$\langle i_t^2 \rangle = \frac{4kTB}{R_{eq}} \tag{4.23}$$

where R_{eq} is a parallel combination of the junction resistance, external load resistance, and input resistance of the following amplifier. The photodiode series resistance (being very small) is neglected.

As a numerical example, if I_d and I_l are equal to 20 nA, R_{eq} is 50 Ω and B is 1 Hz, the mean square values of the shot noise arising from the dark current, and Johnson noise from the equivalent resistance are 6.41×10^{-27} A and 3.31×10^{-22} A, respectively. Notice that the Johnson noise is several orders of magnitude larger than the shot noise.

The signal-to-noise ratio at the output of the photodiode is therefore

$$(S/N)_{power} = \begin{cases} \dfrac{\langle i_{ph}^2 \rangle}{\langle i_q^2 \rangle + \langle i_{db}^2 \rangle + \langle i_{ds}^2 \rangle + \langle i_t^2 \rangle} & (4.24a) \\[2ex] \dfrac{1/2\,(q\eta m\, P_0/h\nu)^2}{2q(I_p + I_d + I_l)B + \dfrac{4kTB}{R_{eq}}} & (4.24b) \end{cases}$$

The minimum optical power required to obtain a given signal-to-noise ratio assuming 100% modulation ($m = 1$) therefore is

$$(P_0)_{min} = \frac{2h\nu B}{\eta}(S/N)\left\{1 + \left[1 + \frac{I_{eq}}{qB\,(S/N)}\right]^{1/2}\right\} \qquad (4.25)$$

where

$$I_{eq} = I_d + I_l + \frac{2kT}{qR_{eq}}$$

In the limit, when $I_{eq}/qB\,(S/N)$ is much less than unity the minimum optical power is determined by the quantum noise associated with the optical signal and is given by

$$(P_0)_{min} = \frac{2h\nu B}{\eta}(S/N) \qquad (4.26)$$

However, when $I_{eq}/qB\,(S/N)$ is much larger than unity, the thermal noise due to R_{eq} becomes dominant. A related figure of merit for photodetectors is the *noise-equivalent power* (NEP). The NEP is the incident rms optical power required to produce a signal-to-noise ratio of 1 in a 1 Hz bandwidth. Under this condition, the NEP is given by [4]:

$$\text{NEP} = \text{rms optical power } (P_0)_{min} \text{ with } (S/N) = 1 \text{ and } B = 1 \text{ Hz}$$

$$= \sqrt{2}\left(\frac{h\nu}{\eta}\right)\left(\frac{I_{eq}}{q}\right)^{1/2} \qquad (4.27)$$

The NEP is expressed in $W \cdot Hz^{1/2}$. To improve the sensitivity of the photodiode, η and R_{eq} should be increased, while I_d and I_l should be decreased. As a numerical example, let us suppose that for a typical InGaAs-InP PIN photodiode, if $\eta = 70\%$ at $\lambda = 1.3$ μm $I_d + I_l = 20$ nA and $R_{eq} = 50$ Ω, the NEP = 24.8 $pW \cdot Hz^{1/2}$.

4.2.8 Reliability

A good measure of the reliability of PIN photodiodes is the increase in the dark current with time and temperature [12–14]. Life tests using thermal overstress at a few hundred degrees Celsius and under normal reverse bias show that the photodiodes typically survive for many thousands of hours. A photodiode is considered to have failed when the dark current increases from an initial value of a few tenths to a few nA. This is equivalent to several hundred years of service life at 70° Celsius [13]. An interesting observation made during the reliability studies is that, by temperature annealing without bias voltage or briefly applying a high reverse bias voltage, the dark current increase largely can be revoked [14].

The radiation harness of photodiodes will be discussed in Section 8.5.3 of Chapter 8.

4.3 PLANAR PHOTOCONDUCTIVE DETECTOR

One of the major obstacles to the successful integration of a field-effect transistor, which is a planar device, and a PIN photodiode, which is a vertical device, is the incompatibility of the device structures and fabrication process. Therefore, a planar photodiode is highly desirable for realizing a simple monolithic integrated optoelectronic circuit. A planar photoconductive photodetector is one such structure [15–21]. A planar photoconductive detector is simply a slab or an epilayer of semiconductor with ohmic or Schottky contacts for electrical connection as illustrated in Figure 4.9.

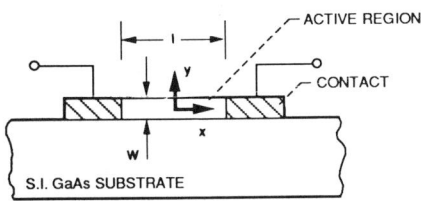

Figure 4.9 Schematic illustrating a lateral photoconductive detector.

4.3.1 Photoconductive Gain and Carrier Lifetime

When intensity-modulated optical illumination of energy greater than the semiconductor bandgap is absorbed, electron-hole pairs are generated. The resulting primary photocurrent is given by [15]:

$$I_0 = q \int_{V_0} G \, dV \qquad (4.28)$$

where the generation rate G is assumed to be uniform over the entire active volume V_0. The number of electron-hole pairs that are generated per second is [15]:

$$\int_{V_0} G \, dV = \frac{\eta P_0}{h\nu} \qquad (4.29)$$

where η is the internal quantum efficiency of the device. The gain of the photodetector Γ is defined as the photocurrent I flowing between the electrodes as a consequence of the primary photocurrent I_0, given by (4.28). Hence,

$$\Gamma = \frac{I}{I_0} = \frac{h\nu I}{q\eta P_0} \qquad (4.30)$$

From (4.29) and (4.30), the gain Γ is

$$\Gamma = \frac{I}{qGV_0} \qquad (4.31)$$

If R_f is the Fresnel reflection at the surface of the photoconductive detector, the total power absorbed in the active region can be written according to (4.5). The number of electron-hole pairs generated per cm³ is $G\tau_{\text{eff}}$, where τ_{eff} is the effective life-time. Therefore, the electron-hole pairs density Δn, which is equal for electrons and holes, is

$$\Delta n = G\tau_{\text{eff}} \qquad (4.32)$$

The corresponding photocurrent density, if none of the generated charge is trapped, is

$$J = q\Delta n v \qquad (4.33(a))$$

where v is the carrier velocity. The dependence of the carrier velocity on the externally applied electric field E for several semiconductors, is shown in Figure 4.10. The photoconductive detector is normally operated at a point on this characteristic where the velocity is a maximum and is denoted as v_{max}. The corresponding electric field is called the *critical electric field* and is denoted as E_c. Hence,

$$J = q\Delta n\,(\mu_n + \mu_p)\,E \qquad (4.33(b))$$

where μ_n and μ_p are the mobilities of electrons and holes, respectively. If A and l are the cross-sectional area and length, respectively, of the photoconductor active region, then

$$V_0 = Al \qquad (4.34)$$

and

$$I = AJ \qquad (4.35)$$

In a practical photoconductive detector, l would be the distance of separation between the fingers of an interdigital structure. The electric field is assumed to be constant along the length of the device, and is equal to

$$E = \frac{v}{l} \qquad (4.36)$$

where V is the externally applied voltage. From (4.31) through (4.36), the gain is

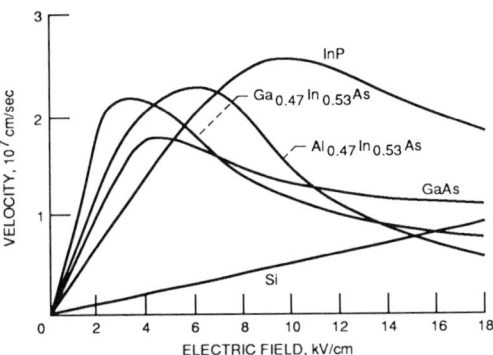

Figure 4.10 Static velocity-field characteristics for a number of semiconductors of interest. The doping level is $N_D = 10^{17}/cm^3$.

$$\Gamma = \tau_{\text{eff}}(\mu_n + \mu_p)\frac{V}{l^2} \tag{4.37}$$

The term τ_{eff} is a combination of three terms given by

$$\frac{1}{\tau_{\text{eff}}} = \frac{1}{\tau_0} + \frac{1}{\tau_c} + \frac{1}{\tau_S} \tag{4.38}$$

where τ_0, τ_c, and τ_S are the lifetimes determined by the volume, contact recombination, and surface recombination, respectively. These lifetimes are a function of the active region geometry, and can be analytically expressed in terms of the surface recombination velocity, S, diffusion lengths, L_n and L_p, and physical dimensions, l and W [15]. However, from a practical standpoint, a photoconductive detector can be categorized under any one of the four cases discussed below, and for which approximate values of τ_{eff} can be calculated.

Case I: Long and thick device for which $l \gg L_n$ or L_p and $W \gg S\tau_0$. The effective lifetime is

$$\tau_{\text{eff}} = \tau_0 \tag{4.39}$$

Case II: Short and thick device for which $l \ll L_n$ or L_p and $W \gg S\tau_0$:

$$\tau_{\text{eff}} = \frac{l^2}{12D} \tag{4.40}$$

where D is equal to D_n or D_p

Case III: Long and thin device for which $l \gg L_n$ or L_p and $W < S\tau_0$:

$$\tau_{\text{eff}} = \frac{W}{2S}\frac{1}{1 + W/2S\tau_0} \tag{4.41}$$

For high surface recombination velocities, $S \gg W/2\tau_0$,

$$\tau_{\text{eff}} = \frac{W}{2S} \tag{4.42}$$

Case IV: Short and thin device for which $l < L_n$ or L_p and $W < S\tau_0$,

$$\tau_{\text{eff}} = \frac{1}{\left(\dfrac{2S}{W} + \dfrac{12D}{l^2}\right)} \tag{4.43}$$

As a numerical example, let us suppose that for a GaAs photoconductive detector W is equal to 1 μm and l is equal to 3 μm, and we wish to calculate τ_{eff}. We know that S is equal to 10^6 cm/s, τ_0 is equal to 2 ns and L_n is equal to 5 μm for n-type GaAs [15]. Based on these parameters, we observe that l is less than L_n and W is less than $S\tau_0$. Hence, the photodetector falls under case IV discussed above, for which τ_{eff} from (4.43) is 27.3 ps.

4.3.2 Gain-bandwidth Product and Rise Time

The bandwidth B is defined as [15]:

$$B = \frac{1}{2\pi \tau_{\text{eff}}} \qquad (4.44)$$

As a numerical example, for the photodetector in the last example, the bandwidth according to (4.44) is 5.84 GHz.

The gain bandwidth product is expressed as [15]:

$$\Gamma B = \frac{1}{2\pi} (\mu_n + \mu_p) \frac{V}{l^2} \qquad (4.45)$$

As a second numerical example, let us suppose that μ_n and μ_p are equal to 7.5 × 10^3 cm^2/V-s and 4.0 × 10^2 cm^2/V-s, respectively for GaAs at 300 K and for an impurity concentration of 10^{14}/cm^3. If the distance of separation between the electrodes is 3 μm and a reverse bias of 5 V is applied, then the gain bandwidth product is 69.9 GHz. If the detector is illuminated by an optical pulse, then the 10 to 90% rise time, assuming the photoconductive detector to behave as an RC circuit, is [15]:

$$\tau_r = \frac{0.36}{B} \qquad (4.46)$$

As a further numerical example, for the photodetector in the last example, the rise time according to (4.46) is 61.7 ps.

The fall time τ_f is assumed to be equal to τ_r if the device does not show storage behavior. The above model is not fully adequate for quantitatively describing photoconductivity, but it is useful to indicate the importance of carrier lifetime, *et cetera*, on the photoresponse characteristic.

4.3.3 Signal-to-Noise Ratio and Noise Equivalent Power

To obtain an expression for the signal-to-noise ratio, the source of optical illumination is assumed to be intensity-modulated by a sinusoidal signal. The resulting intensity-modulated optical signal is given by [4]:

$$P(\omega) = P_0 (1 + m\, e^{j\omega t}) \qquad (4.47)$$

where

P_0 = average optical signal power,
m = modulation index,
ω = modulation frequency.

The rms signal current resulting from the optical signal can be written as [4]:

$$i_p = \frac{q\eta m P_0}{\sqrt{2}\, h\nu} \left[\frac{\tau_0}{t_r}\right] \frac{1}{(1 + \omega^2 \tau_0^2)^{1/2}} \qquad (4.48)$$

where t_r is the transit time, which is equal to $1/v$. If the photodetector dark conductance is G, thermal (Johnson noise) resulting from the conductance G is [4]:

$$i_G^2 = 4\, kTGB \qquad (4.49)$$

where

k = Boltzmann's constant,
T = absolute temperature,
B = bandwidth.

The generation-recombination or shot noise is given by [4]:

$$i_{GR}^2 = \frac{\tau_0}{t_r} \frac{4\, qIB}{(1 + \omega^2 \tau_0^2)} \qquad (4.50)$$

where I is the steady-state, light-induced output current equal to

$$I = \frac{\tau_0}{t_r} \frac{q\eta P_0}{h\nu} \qquad (4.51)$$

The signal-to-noise ratio is defined as [4]:

$$(S/N)_{\text{power}} = \frac{i_p^2}{i_{GR}^2 + i_G^2}$$

$$= \eta m^2 \frac{(P_0/h\nu)}{8B} \left[1 + \frac{kTt_r(1 + \omega^2 \tau_0^2)G}{q\tau_0 I} \right]^{-1} \quad (4.52)$$

The rms optical power is m $P_0/\sqrt{2}$; hence, by setting $B = 1$ Hz and $S/N = 1$ in (4.52), an expression for the NEP is obtained:

$$\text{NEP} = \frac{mP_0}{\sqrt{2}} = \frac{h\nu \, 2^{2.5}}{\eta m} \left[1 + \frac{kT \, t_r (1 + \omega^2 \tau_0^2) G}{q\tau_0 I} \right] \quad (4.53)$$

As a numerical example, let us suppose that

$$\lambda = 0.83 \, \mu\text{m}$$
$$\eta = 60\%$$
$$m = 1$$
$$P_0 = 0.1 \text{ mW}$$
$$t_r = 1/\nu = 3 \times 10^{-4}/10^7 = 30 \text{ ps}$$
$$\tau_0 = 1 \text{ ns}$$
$$f = 10 \text{ GHz}$$
$$G = 0.02 \text{ mho}$$

then the NEP is 1.03 fW·Hz.

4.3.4 Practical Photoconductive Detector Structures

A literature survey [16–22] shows that the device structure of practical photoconductive detectors is of two types. The first type makes use of a simple semi-insulating GaAs substrate. Thus, free electron-hole pairs are generated in the substrate when illuminated optically. The second type has an epilayer a few μm thick of lightly doped semiconductors such as n^- GaAs grown by VPE, MOCVD, or ion implantation on a semi-insulating GaAs substrate. The epilayer absorbs optical radiation at the wavelength of interest, forms the conductive channel of the detector, and minimizes the influence of the semi-insulating substrate which has many traps due to large numbers of deep donors, acceptors, and defects. In both types of detectors a conducting electrode pattern is deposited on the top surface, which is optically illuminated. An interdigital electrode pattern is chosen because it can provide more active area without blockage for illumination. A wide choice of metals or metal alloys is available for the electrodes [18]. However, for low dark current, the electrode metal should result in a Schottky-barrier height of approximately half the band gap of GaAs [18,20]. This aspect will be discussed in greater detail in Section 4.3.5. Figure 4.11 schematically illustrates typical interdigital photoconductive detector structures. The detectors in Figure 4.11(a) and (b) behave like back-to-back Schottky diodes and their typical I-V characteristic is also shown in the inset. The detector in Figure 4.11(c) behaves like a voltage dependent resistor and hence, its I-V characteristic shown in the inset is a straight line (Ohm's law) through the origin. Table 4.2 summarizes the performance of typical photoconductive detectors which are capable of detecting an analog-modulated microwave optical signal.

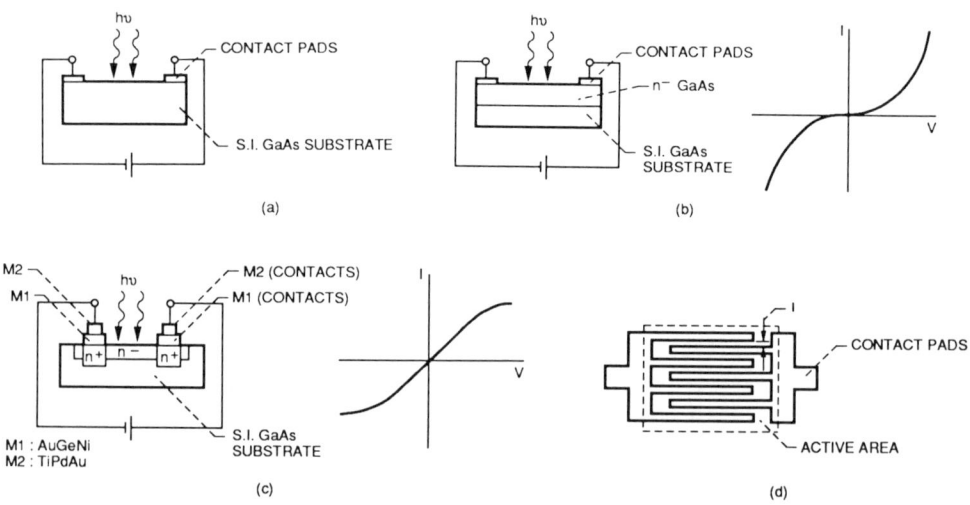

Figure 4.11 Schematic illustrating practical photoconductive detectors on semi-insulating GaAs substrates: (a) semi-insulating substrate with Schottky contacts; (b) n^- GaAs epilayer with Schottky contacts; (c) ion-implanted n^- GaAs epilayer with ohmic contacts; (d) Interdigitated electrode structure of the detector.

4.3.5 Dark Current

The energy band diagram of a metal-semiconductor-metal photoconductive detector under thermal equilibrium is shown in Figure 4.12(a). In this figure, Φ_{Bn}, Φ_{Bp}, and E_F denote the barrier height for electrons, barrier height for holes, and Fermi level, respectively. If we applied a reverse-biased voltage V across the detector and increased it sufficiently beyond reach until the flat band condition is reached, the energy band diagram would appear as shown in Figure 4.12(b). Thermionic injection

Table 4.2
Device Structure and Measured Performance Comparison of Typical Photoconductive Detectors

Device Structure	Substrate	Epilayer	Electrode Metal-Contact	Dark Current	Active Area (μm^2)	Capacitance (pF)	Responsivity (A/W)	Inter-electrode Spacing (μm)	3 dB Bandwidth (GHz)
Fig. 4.11(a)	Si GaAs	—	Cr-Au-Schottky	20 nA at 10 V	100 × 100	0.2	0.5 at 0.81 μm	3	7.0 [17]
Fig. 4.11(b)	Si GaAs	n⁻ GaAs 5 μm ≈$10^{14}/cm^3$	Al-Schottky	1 μA at 10 V	22.5 × 18.0	0.04	0.17 at 0.82 μm	1.5	18.0 [21]
Fig. 4.11(b)	Si GaAs	1 μm Undoped	Al-Schottky	0.8 nA at 4 V	10 × 15	0.012	0.1 at 0.8 μm	0.5	105 [64]

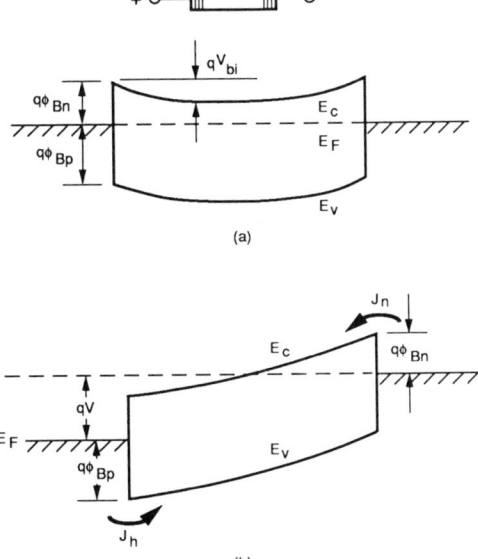

Figure 4.12 Energy-band diagram of a metal-semiconductor-metal photoconductive detector with uniform doped *n*-type semiconductor: (a) under thermal equilibrium; (b) under bias.

of electrons occurs at the reverse-biased contact, while holes are injected at the forward-biased contact. In addition, there is the thermally induced dark current density J_t. Therefore, the total current is the sum of the three constituent currents and is mathematically represented as [18]:

$$J_{total} = J_n + J_h + J_t \qquad (4.54)$$

where

$J_n \qquad = A_n^* T^2 \exp(-q\Phi_{Bn}/kT) \exp(q\Delta\Phi_{Bn}/kT)$
$J_h \qquad = A_p^* T^2 \exp(-q\Phi_{Bp}/kT) \exp(q\Delta\Phi_{Bp}/kT)$
A_n^* and A_p^* = Richardson constants for electrons and holes, respectively,
$\Delta\Phi_{Bn}$ and $\Delta\Phi_{Bp}$ = electron and hole image force barrier height lowering, respectively.

As a numerical example, let us consider a metal-semiconductor-metal photodiode formed by gold and n-type GaAs semiconductor. The Schottky-barrier height is given by

$$\Phi_{Bn} = \Phi_m - \chi = 4.82 - 4.07 = 0.75 \text{ eV}$$

where Φ_m and χ are the metal work function and the semiconductor electron affinity, respectively. The Schottky-barrier potential lowering is determined from

$$\Delta V_B = \left[\frac{qE}{4\pi\varepsilon_0\varepsilon_r}\right]^{1/2}$$

and is equal to 21.05 mV for a typical E of 4×10^6 V/m. The effective Richardson constant for thermionic emission is 1.2×120 A/cm^2/K^2 for n-type GaAs under high field [4]. Hence,

$$J_n = 1.2 \times 120 \times (300)^2 \exp(-0.75/26 \times 10^{-3})$$
$$\cdot \exp(21.05 \times 10^{-3}/26 \times 10^{-3})$$
$$= 8.64 \text{ }\mu\text{A/cm}^2$$

If the device is 3 μm long, 100 μm wide, and 2 μm high, the cross-sectional area is 200×10^{-8} cm^2, and hence the dark current due to electron injection is 17.3 pA.

Equation (4.54) clearly shows that the electron and hole injection rates are governed exponentially by their respective barrier heights. Assuming that the image force lowering is negligible in the semi-insulating substrate, the sum of the currents will be minimized when the barrier heights fall near the mid band gap of the semiconductor [18,20]. The respective barrier height of tungsten and tungsten silicide are 0.67 eV and 0.7 eV, roughly half the band gap of GaAs. Hence, photoconductive detectors that use these metals instead of traditional metals for contacts on GaAs have comparably lower dark currents as shown in Table 4.2. InGaAs has superior mobility compared to other materials [23], but the Schottky-barrier height between a metal and n-type InGaAs is on the order of 0.3 eV [24]. Hence, photoconductive detectors fabricated with InGaAs as an epilayer material have resulted in an unacceptably high dark current of several mA at a few volts bias [25]. The above discussions suggest that the low Schottky-barrier is one of the causes for a large value of the dark current.

Finally, measurements [26] show that the dark current increases exponentially with temperature at a fixed bias which is in accordance with (4.54). However, the noise power at a fixed temperature and bias decreases as the frequency increases, which is in accordance with (4.50).

4.3.6 Lumped Element Equivalent Circuit Model

In most practical photodetectors, the carrier transit time is limited by the parasitic resistance and reactances. Hence, by tailoring the device resistance, inductance, and capacitance, we are able to tune out the reactance and to impedance match the resistance to a load over a broad band of frequencies. However, in the case of a practical photoconductive detector, the carrier transit time is not limited by the RC and LR time constants because of a long tail observed in the measured detector's optical impulse response [16]. The tail is thought to be due to the trapping of photogenerated minority carriers (holes) in the Cr doped semi-insulating GaAs substrate. In such a situation, no advantage is gained by tuning out the detector reactance because the bandwidth is limited by the hole-emission time. Trapping of carriers will be explained in Section 10.10 of Chapter 10. In general, as has been experimentally observed, at a fixed frequency, if the bias voltage across the photoconductive detector is gradually varied from zero to a large value, the input reflection coefficient locus includes the center of the Smith chart [22], as shown in Figure 4.13(a). Thus, there

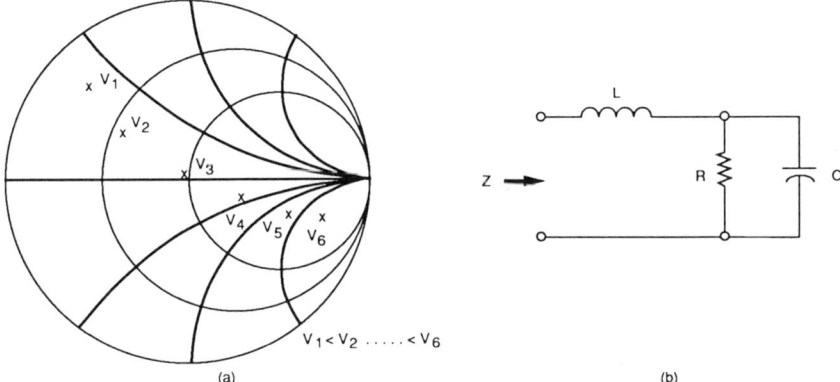

Figure 4.13 (a) Input reflection coefficient, S_{11}, of the detector as a function of the bias voltage; (b) small-signal, lumped-element equivalent circuit model (from [22], p. 118. Reprinted with permission).

is a bias voltage for which the input impedance is purely resistive. A lumped-element equivalent circuit model [22], shown in Figure 4.13(b), is used to explain this result. The model consists of a series inductance L, which represents the bond wire inductance and is independent of the bias, a voltage dependent resistor R, which takes into account the changing active channel width with bias, and finally a capacitance C, which is due to the interdigitated electrodes. The capacitance C varies with bias because the electric field in the device changes as the active channel width varies.

The impedance looking into the detector model is therefore given by

$$Z = \frac{R}{1 + (\omega C R)^2} + j\left[\omega L - \frac{\omega C R^2}{1 + (\omega C R)^2}\right] \quad (4.55)$$

where ω is the operating angular frequency. At low bias voltages, the channel resistance is small and effectively shorts the capacitive reactance. That is, $R \ll 1/\omega C$, then (4.55) reduces to

$$Z = R + j\omega L \quad (4.56)$$

The model therefore correctly predicts the observed behavior. At high bias voltages, the channel resistance is large and no longer shorts the device capacitance. Further, the capacitive reactance is greater than the inductive reactance. That is, $R \gg 1/\omega C$ and $1/\omega C \gg \omega L$ or $\omega^2 \ll 1/LC$, and (4.55) simplifies to

$$Z = R/(\omega C R)^2 + 1/(j\omega C) \quad (4.57)$$

This equation represents a resistor R in parallel with a capacitor of value C. Again, the model correctly predicts the observed behavior. The point at which the input impedance is purely resistive can be determined from (4.55). If $(\omega C R)^2 \ll 1$, the input reactance is zero for

$$R = \left(\frac{L}{C}\right)^{1/2} \quad (4.58)$$

and hence,

$$Z = \left(\frac{L}{C}\right)^{1/2} \quad (4.59)$$

The zero reactance condition is valid up to frequencies determined by $(\omega C R)^2 \ll 1$.

As an experimental check [22], the measured S_{11} (real and imaginary) for a practical photoconductive detector at a frequency of 0.5 GHz, and a bias voltage of

0 V indicates L and R values of 3 nH and 9 Ω, respectively. The resistance value is verified by comparing it to that predicted from the slope of the I-V characteristic at the origin, which for the above detector is 10 Ω. At a higher bias voltage of 3 V, the measured S_{11} (real and imaginary) at a frequency of 2 GHz shows a C and R value of 0.73 pF and 400 Ω, respectively. Finally, based on the above L and C, the computed zero reactance resistance from (4.58) is 63 Ω, which is in good agreement with the measured S_{11} (real) of 58 Ω. The expected bandwidth computed from C and R is 3.4 GHz.

4.4 SCHOTTKY-BARRIER PHOTODIODES

A typical planar Schottky-barrier photodiode consists of a semitransparent metal film, which is a few tens of microns in diameter and a few hundred angstroms thick deposited on an epilayer [27–35]. The n^- GaAs epilayer is a few microns thick and grown on top of an n^+ GaAs substrate. Further, the metal film is in contact with an offset metal bonding pad for connection to an external circuit. The film surface may be protected by a dielectric film, a few thousand angstroms thick, which also acts as an antireflection coating. An ohmic contact on the backside of the substrate completes the device. A cross-sectional view of the device is shown in Figure 4.14. Under normal operating conditions a reverse-biased voltage is applied between the bonding pad and the backside contact.

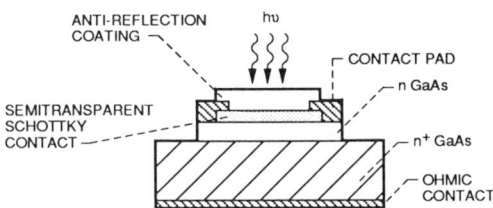

Figure 4.14 Cross section of a GaAs Schottky-barrier photodiode.

There are several advantages in using a Schottky-barrier photodiode. First, a feature common to both PIN and Schottky-barrier diodes is that the absorption layer thickness can be engineered to obtain the optimum compromise between external quantum efficiency and detector bandwidth [27]. Second, the absence of the p^+ region in a Schottky diode eliminates the effect of minority carriers with low mobility [28]. Third, the fabrication processes for Schottky-barrier photodiodes and MESFETs are compatible, and therefore easily integrated. The disadvantages of the Schottky-barrier diode are its high series resistance and low efficiencies, which are associated with the semitransparent metal layer [28]. Another disadvantage of this device, as compared with PIN diodes, is its low responsivity due to its small area. However, the small area results in low parasitics, and hence large bandwidth.

4.4.1 Quantum Efficiency

The energy-band diagram of a Schottky-barrier photodiode under reverse bias is illustrated in Figure 4.15. The metal film is made thin enough to allow a substantial amount of light to reach the semiconductor. There are three photocurrent components. Light with energy $h\nu$ greater than the barrier $q\Phi_{Bn}$ can be absorbed in the metal and excite electrons over the barrier into the semiconductor (1 in Figure 4.15). Light entering the semiconductor is mainly absorbed in the depletion region (2 in Figure 4.15). The light that passes beyond the depletion region is absorbed in the neutral region, creating electron-hole pairs (3 in Figure 4.15). The holes must diffuse to the depletion region edge to be collected.

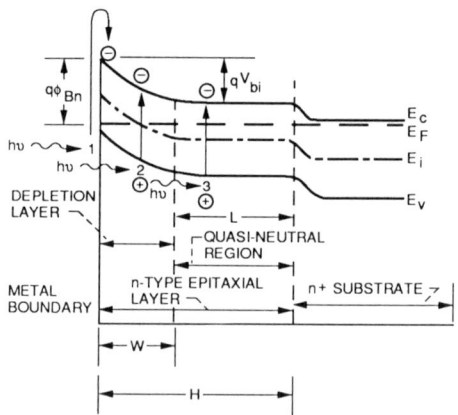

Figure 4.15 Energy-band diagram of an epitaxial Schottky-barrier diode.

The excitation of carriers from the metal into the semiconductors is very small, and therefore can be neglected. In the depletion region, the high field will sweep out the photogenerated carriers before they can recombine, leading to a photocurrent density [4]:

$$J_{dr} = q\alpha T \frac{P_{opt}}{h\nu}[1 - \exp(-\alpha W)] \qquad (4.60)$$

where

T = transmission coefficient of the metal at the wavelength of operation;
P_{opt} = incident optical power density, W/cm^2;
W = width of the depletion region.

If the back contact is ohmic and the thickness of the depletion region is less than that of the epilayer, the photocurrent density due to the hole collected in the neutral region is [4,27]:

$$J_p = \left[\frac{q \alpha P_{opt} T L_p}{(\alpha^2 L_p^2 - 1) h\nu}\right] \exp(-\alpha W) \left[\alpha L_p - \frac{(A-B)}{C}\right] \quad (4.61)$$

where

L = $H - W$ = difference in the thickness between the epilayer and the depletion region,
A = $\cosh(L/L_p)$,
B = $\exp(-\alpha L)$,
C = $\sinh(L/L_p)$,
L_p = diffusion length = $(D_p \tau_p)^{1/2}$.

Thus, the total photocurrent density is the sum of (4.60) and (4.61), and is equal to

$$J_{ph} = q \alpha T \frac{P_{opt}}{h\nu} (1 - \exp(-\alpha W))$$
$$+ \left[\frac{q \alpha P_{opt} T L_p}{(\alpha^2 L_p^2 - 1) h\nu}\right] \exp(-\alpha W) \left[\alpha L_p - \frac{(A-B)}{C}\right] \quad (4.62)$$

The internal quantum efficiency η is the number of electron-hole pairs generated per incident photon of energy $h\nu$ and is given by

$$\eta = \frac{J_{ph}/q}{P_{opt}/h\nu} \quad (4.63)$$

$$\eta = T\left\{1 + \frac{\exp(-\alpha W)}{(\alpha^2 L_p^2 - 1)}\left[1 - L_p \frac{(A-B)}{C}\right]\right\} \quad (4.64)$$

As a numerical example, let us suppose that a GaAs Schottky-barrier photodiode, similar to that shown in Figure 4.14, has the following typical parameters:

wavelength, $\lambda = 0.85 \ \mu m$

semitransparent gold film thickness = 100 Å

transmittance of the gold film, $T = 55\%$ [36]

absorption coefficient, $\alpha = 10^4/\text{cm}$

depletion layer width, $W = 0.5$ μm

n-type epilayer doping density = $10^{15}/\text{cm}^3$

epilayer thickness, $H = 1.5$ μm, and hence, $L = 1.0$ μm

diffusion length, $L_p = 6$ μm [37]

The computed quantum efficiency from (4.64) is 56%.

4.4.2 Response Speed and 3 dB Bandwidth

The response speed is determined primarily by the transit or drift time in the depletion region, diffusion time in the neutral region, and *RC* time constant required to discharge the junction capacitance through a parasitic resistance. In general the transit time and the diffusion time can be made negligible compared to the *RC* time constant by optimizing the device configuration. Thus, the response speed can be estimated from the *RC* time constant and is expressed as

$$t_{RC} = (R_S + R_L) C_j \qquad (4.65)$$

where

R_S = series resistance composed of the spreading resistance and the sheet resistance of the thin epilayer,

R_L = load resistance.

The corresponding 3 dB upper cut-off frequency is

$$f_C = \frac{1}{2\pi(R_S + R_L) C_j} \qquad (4.66)$$

As a numerical example, let us suppose that R_S, R_L, and C_j are equal to 2 Ω, 50 Ω, and 0.2 pF, respectively for a typical photodiode. Then, the 3 dB upper cut-off frequency is 15.3 GHz.

4.4.3 Dark Current

The dark current depends strongly on the Schottky-barrier height. The total dark current for a Schottky-barrier photodiode consists of the thermionic-emission current, the recombination current via traps in the depletion region, the tunneling current due to carriers tunneling across the bandgap, and the leakage current or interface current due to traps at the metal-semiconductor interface [27]. Most practical diodes have an epilayer with a doping concentration less than $10^{17}/cm^3$, which is considered to be low. Hence, the tunneling current can be neglected. Further, for a good diode, surface leakage current is negligible. Hence, the total dark current is composed of the thermionic-emission current over the Schottky-barrier and the generation-recombination current in the depletion region.

The dark current density as described by thermionic-emission theory is [27]:

$$J = A^*T^2 \exp(-q\ \Phi_{Bn}/kT)[\exp(qV/nkT) - 1] \quad (4.67)$$

$$= J_S [\exp(qV/nkT) - 1] \quad (4.68)$$

where

- n = ideality factor, which is close to unity at low doping and high temperature;
- A^* = effective Richardson constant;
- Φ_{Bn} = barrier height;
- V = externally applied voltage across the diode.

The generation-recombination current density through the midgap traps in the depletion region is given by [27]:

$$J_{gr} = J_{r0} [\exp(qV/2kT) - 1] \quad (4.69)$$

where

- $J_{r0} = qn_iW/2\tau_0$
- τ_0 = minority carrier lifetime in the depletion region,
- n_i = intrinsic carrier concentration.

The total dark current is therefore the sum of (4.68) and (4.69). As a numerical example, for a GaAs Schottky-barrier photodiode of diameter 100 μm and with gold contacts, the dark current at a reverse bias of -5 V and room temperature calculated from (4.67) is -0.3 nA. The effective Richardson constant and ideality factor are taken as 1.2×120 A/cm^2/K^2 and unity, respectively. Further, if n_i, W, and τ_0 are equal to 1.79×10^6/cm^3, 0.5 μm, and 10^{-9} s, respectively, the generation-recombination current calculated from (4.69) is -0.56 pA.

4.4.4 Practical Schottky-barrier Photodiodes

Practical planar Schottky-barrier photodiodes have been mostly fabricated on semi-insulating GaAs substrates. The epilayer, or absorption layer, is n^- GaAs, a few microns thick. Thus, the wavelength range over which the diode can operate is from about 0.6 to 0.83 μm. A wide choice of metals is available for the semitransparent Schottky contact. However, these metal films result in poor transmittance of light, and hence lower photoresponsivities. One technique to overcome this problem is to use indium tin oxide (ITO) film to form the Schottky contact [29]. The transmittance of ITO is about 93% in the wavelength range 0.4 μm to 0.9 μm. The Schottky-barrier height between GaAs-ITO interface is about 0.9 eV, which is about the same as that for a GaAs-gold interface.

For applications in the wavelength range of 1.3 to 1.6 μm a Schottky contact photodiode on p-type GaInAs has been demonstrated. Worth noting at this instance is that the Schottky-barrier between a metal and an n-type $In_xGa_{1-x}As$ is composition dependent and is on the order of 0.3 eV for the lattice-matched value x equal to 0.53. This value is considered to be low [24]. However, in the case of p-type material, it is on the order of 0.7 eV [30], which is adequate for most practical purposes. Table 4.3 summarizes the performance of typical Schottky-barrier photodiodes capable of detecting an analog-modulated microwave optical signal.

Table 4.3
Device Structure and Measured Performance Comparison of Typical Planar Schottky-Barrier Photodiodes

Device Structure	Dimensions	Schottky Metalization	R (Ω)	C (fF)	Responsivity (A/W)	External Quantum Efficiency	Leakage Current	−3 db Bandwidth (GHz)	Reference
p − GaInAs- p^+ − GaInAs	18 μm (diameter)	Ni	190	25	—	19 at 1.27 μm	0.5 μA at −6 V	>23.34	[31]
η − GaAs- η^+ − GaInAs	5 μm × 5 μm	ITO ($In_{0.9}Sn_{0.1}O$)	—	≤30	0.2	>25 at 0.82 μm	<1 nA at −5 V	>110	[28,34]

4.4.5 Lumped Element Equivalent Circuit Model

A lumped-element equivalent circuit model [35] for a Schottky-barrier photodiode is shown in Figure 4.16. The capacitance C is primarily due to the depletion layer and is typically a few tenths of a pF. The inductance L is primarily due to the bonding wire, and is typically a few tenths of an nH. The series resistance, R, is due to the sheet resistivity of the Schottky-barrier layer and is typically a few Ω. The current

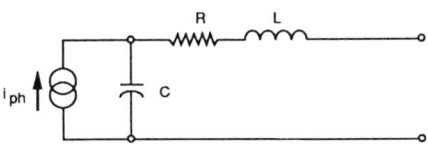

Figure 4.16 Lumped-element equivalent circuit model.

source of magnitude i_{ph} in parallel with C represents the rms photogenerated signal current.

4.5 AVALANCHE PHOTODIODES

Avalanche photodiodes internally multiply the primary signal photocurrent before it enters the input circuitry of the following amplifier. This increases receiver sensitivity, as the photocurrent is multiplied before encountering the thermal noise associated with the receiver circuit. For carrier multiplication to occur, the photogenerated carriers must traverse a region where a very high electric field is present. In this high-field region, a photogenerated electron or hole can gain enough energy so that it ionizes bound electrons in the valence band upon colliding with them. This carrier multiplication is known as *impact ionization*. The newly created carriers are also accelerated by the high electric field, thus gaining enough energy to cause further impact ionization. This phenomenon is the avalanche effect.

Silicon avalanche photodiode structures suitable for less than 1 μm wavelength have been discussed elsewhere [4]. In this section, we discuss an InGaAs avalanche photodiode structure suitable for the long wavelength region. The initial long wavelength APDs suffered from excessive dark current, which was attributed to tunneling of carriers at the high electric fields required for avalanche gain [38]. As has been demonstrated, this tunneling component of the dark current can be eliminated with a structure that provides "separate" regions for "absorption" and "multiplication" [38]. This type of device is designated as the SAM-APD. Unfortunately, the bandwidth of the SAM-APD is degraded by charge accumulation at the valence band discontinuity at the heterostructure interface [38]. Recently, it has been demonstrated that by growing an intermediate band gap "grading" layer between the wide-bandgap multiplication layer and the narrow-band-gap absorbing layer the above problem can be eliminated [38,39]. This type of device is designated as the SAGM-APD. This device simultaneously exhibits high quantum efficiency, low dark current, large bandwidth, and avalanche gain [39].

4.5.1 Separate Absorption Grading and Multiplication-Avalanche Photodiode

The schematic cross section of the back illuminated InP-InGaAsP-InGaAs SAGM-APD [38] is shown in Figure 4.17. The structure consists of five lattice matched layers. The first layer is a heavily doped *p*-type InP buffer layer, which is a few microns thick and is grown on top of an InP substrate. This is followed by two lightly doped *n*-type layers (shown as a single layer in Figure 4.17), which form the multiplication region. The combined thickness of this layer is in the range of 1 to 2 μm. Next, a layer of *n*-type InGaAsP is grown, which is typically a few tenths of a micron in thickness and forms the grading layer. Finally, a lightly doped *n*-type InGaAs is grown, which is several microns thick and forms the absorbing layer.

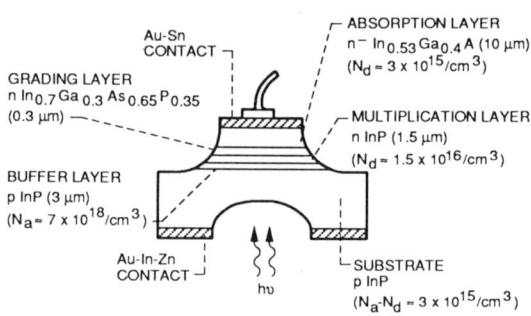

Figure 4.17 Schematic cross section of InP-InGaAsP-InGaAs SAGM structure APD (after [38], p. 118).

A one-dimensional electric field profile of the above device when biased beyond punch through is shown in Figure 4.18. Because the multiplication, grading and absorption regions of the APD are not coincident, the primary holes have an additional drift path which can add to their transit time. Further, the secondary electrons must also traverse the entire depletion region before being collected. Because the latter distance is greater, the transit time of the secondary electrons can be the dominant-bandwidth-limiting effect. The width of the absorption layer can be optimized in a manner similar to that of a PIN diode as explained in Section 4.2.2.

4.5.2 Low Frequency Gain and Responsivity

The photocurrent density through the diode is the low frequency gain M_0 times the number of photogenerated electron-hole pairs and can be written in a manner similar to (4.7):

$$J = \frac{\eta q P_0}{h\nu A} M_0 \left[1 - \exp(-\alpha W)\right] (1 - R_f) \qquad (4.70)$$

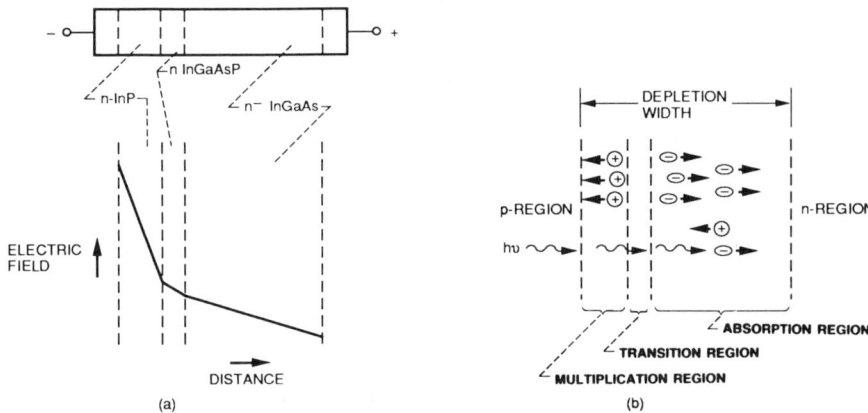

Figure 4.18 (a) One-dimensional electric-field profile when the SAGM-APD is biased beyond punch-through; (b) schematic illustrating the generation of primary and secondary electrons and holes in the various regions of the SAGM-APD.

The low-frequency gain, M_0 can be expressed as a function of the bias voltage as follows [8]:

$$M_0 = \frac{1}{1 - \left(\dfrac{V_r}{V_b}\right)^2} \quad (4.71)$$

where

V_r = reverse bias voltage,
V_b = breakdown voltage.

The responsivity R can be written analogous to (4.15) as

$$R = \frac{\eta q M_0}{h\nu} \quad (4.72)$$

4.5.3 Frequency Response of the APD

The frequency dependence of the gain $M(\omega)$ is given by the expression [40]:

$$\frac{M(\omega)}{M_0} = \frac{\sin(\omega W/2v)}{(\omega W/2v)} D_1 D_2 D_3 \qquad (4.73)$$

where

$$D_1 = \frac{1}{[1 + (\omega RC)^2]^{1/2}}$$

$$D_2 = \frac{1}{[1 + (\omega/e_h)^2]^{1/2}}$$

$$D_3 = \frac{1}{[1 + (\omega \tau_m M_0)^2]^{1/2}}$$

Further, v is the saturated carrier drift velocity (assumed equal for holes and electrons), R is the total series resistance (diode plus load), C is the capacitance, $M_0 \tau_m$ is the avalanche build-up time, and e_h is the emission rate for holes trapped at the heterointerfaces.

The lumped element equivalent circuit model of an avalanche photodiode is similar to that shown in Figure 4.16. Typical element values are indicated in the numerical examples discussed below.

As a numerical example, let us suppose that a typical avalanche photodiode has the following parameters [39]:

$$C = 0.1 \text{ pF}$$
$$R_L + R_S = 60 \text{ }\Omega$$
$$v_n = v_p = v = 5 \times 10^6 \text{ cm/s}$$
$$1/e_h = 4 \times 10^{-12} \text{ s}$$
$$M_0 f \text{ (gain-bandwidth product)} = 1/2\pi \tau_m = 60 \text{ GHz}$$
$$\alpha = 1.15/\mu\text{m}$$
$$W = 2.7 \text{ }\mu\text{m}$$
$$\omega = 2\pi f$$

The normalized gain $M(\omega)/M_0$ computed from (4.72) is then approximately -3 dB at a frequency of 18 GHz. Because the gain bandwidth product is 60 GHz and f (3 dB) is 18 GHz, M_0 is found to be approximately 3.

4.5.4 Signal-to-Noise Ratio and Noise Equivalent Power

Analogous to (4.24b), the S/N for an avalanche diode can be written as follows:

$$\frac{S}{N} = \frac{(1/2)\,(q\,\eta\,m\,P_0/h\nu)^2\,M_0^2}{2\,q\,(I_p + I_d)\,M_0^2\,F(M_0)\,B + 2qI_1B + \dfrac{4kTB}{R_L}} \tag{4.74}$$

where $F(M_0)$ is the noise factor and can be approximated as M_0^x. Because the noise figure $F(M_0)$ increases with M_0, there exists an optimum value of M_0 that maximizes the S/N. The optimum M_0 for $m = 1$ is given by [3]:

$$(M_{\text{opt}})^{x+2} = \frac{2qI_1 + 4kT/R_L}{xq(I_p + I_d)} \tag{4.75}$$

Notice that if $F(M_0) = M_0 = 1$, (4.74) reduces to (4.24b).

As a numerical example, let us assume the following set of parameters for an avalanche photodiode:

$$I_1 = 0.2 \text{ nA}$$
$$I_d = 0.3 \text{ nA}$$
$$I_p = 0.050 \text{ mA}$$
$$R_L = 60 \text{ }\Omega$$
$$x = 0.50 \text{ for InGaAs APD}$$

The M_{opt} as calculated from (4.75) is 5. Finally, analogous to (4.27), the noise-equivalent power is given by

$$\text{NEP} = \sqrt{2}\left[\frac{h\nu}{\eta}\right]\left[\frac{I_{eq}}{q\,F(M_0)^2}\right]^{1/2}$$

where

$$I_{eq} = I_d\,F(M_0) + \frac{2kT}{q\,R_L\,M_0^2}$$

and the contribution from I_1 is neglected.

As a numerical example, for the APD above, let us suppose that η is equal to 70% at 1.3 μm wavelength, then

$$F(M_0) = M_0^x = 3^{0.5} = 1.73$$

and the NEP is 4.2 pW·Hz$^{1/2}$. Notice that, when compared with a PIN diode, the avalanche gain can substantially reduce the NEP.

4.6 COMPARISON OF PHOTODETECTORS

In the case of photoconductive detectors, the gain is proportional to the ratio of the hole lifetime to the electron transit time. Because the hole lifetime is greater than the electron transit time, the holes continue to drift, even after the end of the optical pulse. This increases the detector response time. Because the response time and the bandwidth are inversely proportional, any increase in the gain is at the expense of the bandwidth. The signal-to-noise ratio increases with decreasing dark conductance G or increasing gain. However, any increase in the gain to improve sensitivity decreases the frequency response.

In the case of PIN photodiodes, the absence of gain results in a gain-bandwidth product that is equal to the bandwidth itself. The bandwidth is usually limited by the RC time constant of the device. The principal source of noise is the shot noise in the depletion region of the reverse-biased p-n junction. The shot noise is generally several orders of magnitude smaller than the Johnson noise associated with the channel resistance of the photoconductive detectors. Hence, PIN diodes are generally more sensitive than photoconductive detectors.

In the case of an APD, there is an optimal gain for which the S/N is maximized. Hence, in the presence of the gain, the bandwidth is reduced because time is taken for an avalanche to build up.

A more useful comparison of these detectors can be made if we consider how their choice influences the performance of an RF fiber optic link. One figure of merit by which to compare the devices is to calculate the distance between repeaters at a fixed data rate. Figure 4.19 shows the repeater spacing *versus* the bit rate for the three types of photodetectors. At bit rates below 4 Gb/s, the APD offers the best performance [41].

4.7 FIELD-EFFECT TRANSISTORS

Several investigators have studied the effect of optical illumination on the dc and microwave characteristics of metal semiconductor field-effect transistors [42–46], and the high electron mobility transistors [47–49]. These devices have several advantages when used as photodetectors. First, the detected signal will be internally amplified through the device transconductance. Second, there is an improvement in the output signal-to-noise in the case of HEMTs. This is because the HEMT, being a heterojunction device, has an active channel in which the mobility of the electrons

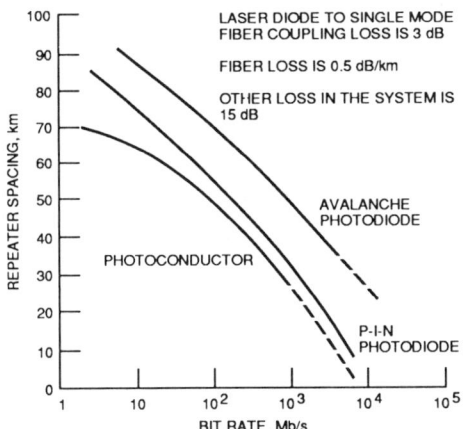

Figure 4.19 Computed repeater spacing *versus* bit rate (from [41], p. 119. Reprinted with permission).

is very high. The high electron mobility is responsible for the high velocity of the electrons in the active channel. Because the velocity is directly related to the unity current gain cut-off frequency, f_t, HEMTs display higher f_t than conventional MESFETs. Also, the noise figure and f_t are inversely related through the Fukui's equation [50]. Hence, a high f_t results in a lower device noise figure. Third, there is a possibility of using materials with high electron mobility, such as AlGaAs, InGaAs, or InGaAsP, to form the heterostructure in a HEMT. The band gap of these materials can be tailored so as to be sensitive to a particular optical wavelength, thereby improving the responsivity of the device. Fourth, as both FETs and optoelectronic devices use the same material and fabrication technology, these devices can be integrated on a single wafer [51].

4.7.1 Device Structure

A typical GaAs MESFET consists of a thin (0.2 μm), highly doped, ($10^{17}/cm^3$) n-type GaAs epitaxial layer grown on a semi-insulating GaAs substrate, as illustrated in Figure 4.20. In most practical devices, aluminum has been used to form the gate Schottky contact and Au/Ge eutectic has been used to form the source and drain ohmic contacts.

In a MESET, the channel is a uniformly doped layer of bulk semiconductor. Therefore, both electrons and holes share the same space and interact via their Coulomb potentials. The effect of the Coulomb interaction on carrier transport is known as *ionized impurity scattering*. The impurity scattering lowers both the electron mobility and saturation velocity. As an example in pure GaAs, the peak velocity (at 300 K)

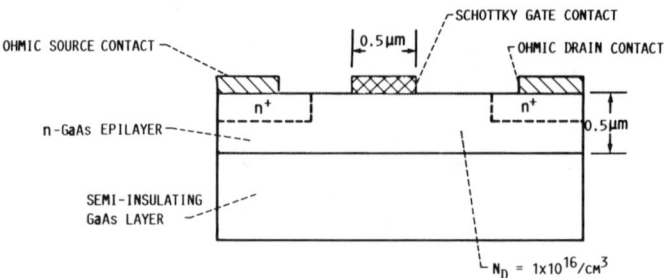

Figure 4.20 Schematic cross section of a typical GaAs MESFET.

decreases from 2.1×10^7 cm/s to 1.8×10^7 cm/s for a doping level of $1 \times 10^{17}/$cm^3.

Figure 4.21 shows a typical superlattice formed from multiple heterojunctions. As seen in the figure, only the larger band-gap AlGaAs material is doped with donor. The heterojunction lineup is such that the energy of the free electrons in the AlGaAs is higher than in the adjacent GaAs. The electrons initially introduced into the AlGaAs diffuse into the lower energy GaAs layer where they are confined due to the energy barrier at the heterointerface in a so-called *two-dimensional electron gas* (2DEG). Sheet electron densities as high as $10^{12}/$cm^2 may be obtained at a single interface. This technique of separating the electrons from the donors greatly reduces the ionized impurity scattering. The transport in the 2DEG approaches that of undoped bulk GaAs. Typically at 300 K, low-field electron mobilities in the range of 8500–9000 cm^2/V·s have been achieved by this technique as compared to about 4000 cm^2/V·s in a GaAs FET channel.

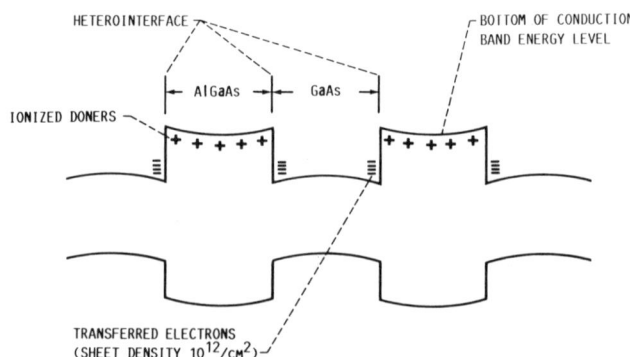

Figure 4.21 Multiple interface AlGaAs/GaAs heterostructure.

To overcome the limitations imposed by ionized impurity scattering in MESFETs, the electrons must be physically separated from the donors. This can be accomplished by using heterostructures in which there is an abrupt discontinuity in the conduction bands, and the valence bands at the heterojunction. This desired feature has been incorporated in a HEMT.

A typical structure of a AlGaAs-GaAs HEMT [52] is depicted in Figure 4.22. The wafer is a semi-insulating Cr-doped GaAs, typically 25 mils thick. The epitaxial layers, which are grown sequentially by using the molecular beam epitaxy process are as follows: an undoped GaAs buffer layer typically about 1 μm thick; a lattice-matched and undoped layer of $Al_{0.3}Ga_{0.7}As$, typically 30 Å thick; a doped (2×10^{18}/cm^3) layer of $Al_{0.3}Ga_{0.7}As$, typically 450 Å thick; a lattice-matched and doped (2.8×10^{18}/cm^3) GaAs layer, typically 100 Å thick. The purpose of the GaAs buffer layer is to move the heterojunction away from the substrate, which may contain traps and other impurities and also to provide an atomically smooth interface on which to

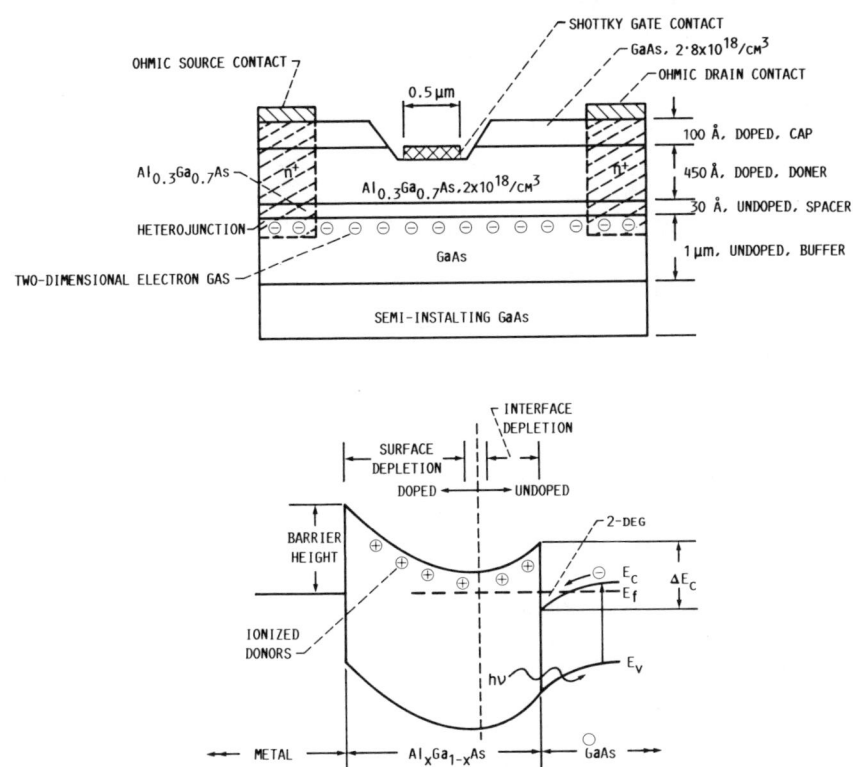

Figure 4.22 Schematic cross section of a typical AlGaAs/GaAs on GaAs HEMT and the energy band diagram.

grow the heterojunction. Further, the GaAs buffer layer is undoped to allow the electrons to travel with maximum mobility unimpeded by dopants. The higher bandgap AlGaAs layer is highly doped so as to provide free electrons which diffuse into the lower band-gap GaAs layer, where they are retained by the energy barrier and form the 2DEG layer, as explained earlier. In addition, an undoped AlGaAs layer is provided at the heterointerface to move the doped AlGaAs layer away from the 2DEG layer, and further improve the mobility of the electrons. Finally, a GaAs cap layer is added because of the difficulty in providing a good electrical and metallurgical contact to AlGaAs.

To control and modulate the 2DEG, a Schottky-barrier gate electrode is placed on the doped AlGaAs layer. The charge in the doped AlGaAs is depleted at the heterojunction by electron diffusion into GaAs. Charge is also depleted from the surface by the built-in voltage of the gate. To avoid conduction through AlGaAs, which has inferior charge transport properties, the two depletion regions must overlap as shown in the energy band diagram in Figure 4.22. In devices normally in the on-state, the gate built-in voltage is just sufficient to extend the surface depletion layer to the interface depletion. In this depletion mode device, application of negative bias to the gate will extend the gate depletion region to the interface raising the barrier to electron flow, thereby pinching off the drain-to-source current. However, in devices normally in the off-state, the thickness of the doped AlGaAs layer under the gate is small, and therefore the gate built-in voltage depletes the doped AlGaAs layer, overcomes the built-in potential at the heterointerface, and depletes the 2DEG. When a positive bias voltage greater than the threshold voltage V_{th} is applied to the gate, electrons accumulate at the heterojunction interface and form a 2DEG, thus turning on the device. Devices which are normally in the on-state are used in low-noise microwave amplifiers, while those which are normally in the off-state are used in digital circuits.

To increase the transconductance, g_m and the unity current gain cut-off frequency, f_t for low-noise amplification at millimeter-wave frequencies, both the mobility and the sheet current density of the two-dimensional electron gas in the channel are to be increased. This can be accomplished by the use of different material systems. Figure 4.23 illustrates two FET device structures based on AlGaAs-InGaAs [53] and AlInAs-GaInAs [54,55] material systems which not only operate at higher frequencies, but also have potential for optoelectronic integration.

In the device shown in Figure 4.23(a), a mole fraction of 0.3 and 0.25 for aluminum and indium, respectively, resulted in a conduction band discontinuity ΔE_c of 0.45 eV. This is higher than the ΔE_c of 0.32 eV in an $Al_{0.3}Ga_{0.7}As$-GaAs material system. Consequently a higher sheet charge density on the order of $2.4 \times 10^{12}/cm^2$ is achieved. Field-effect transistors with gate length and width of 0.2 and 140 μm, respectively, fabricated from AlGaAs-InGaAs material systems exhibited an extrinsic g_m of 550 mS/mm of gate width with corresponding f_t of about 122 GHz [53]. These results compare favorably with the 110 GHz f_t of 0.1 μm gate length AlGaAs-GaAs HEMT [56].

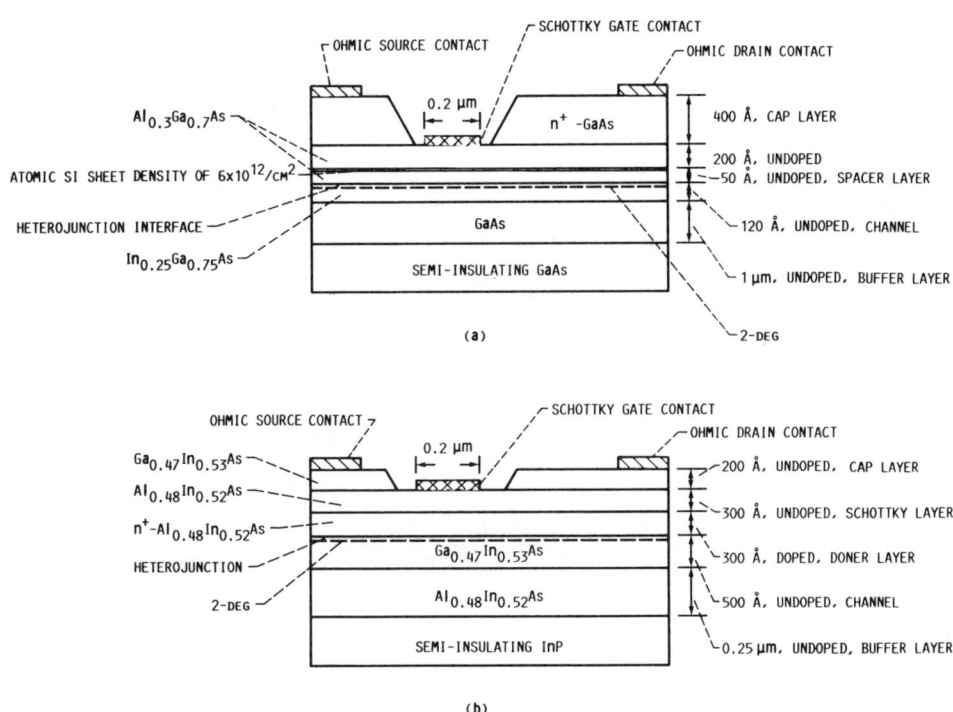

Figure 4.23 (a) Schematic cross section of a typical $Al_{0.3}Ga_{0.7}As/In_{0.25}Ga_{0.75}As$ on GaAs pseudomorphic HEMT; (b) schematic cross section of a typical $Al_{0.48}In_{0.52}As/Ga_{0.47}In_{0.53}As$ on InP HEMT.

In the device shown in Figure 4.23(b), a large conduction band discontinuity ΔE_c between lattice-matched $Al_{0.48}In_{0.52}As$ and $Ga_{0.47}In_{0.57}As$, coupled with the high doping efficiency of Si in AlInAs, resulted in a large sheet charge density on the ratio of $3 \times 10^{12}/cm^2$. Further, the mobility of the electrons in the $Ga_{0.47}In_{0.53}As$ at 300 K is greater than 10000 cm^2/V · s. FETs with gate lengths of 0.2 μm and widths of 50 μm exhibited extrinsic g_m of 800 mS/mm of gate width, and an f_t of about 120 GHz [54].

4.7.2 FET dc Characteristics Under Optical Illumination

When a MESFET or a HEMT device is optically illuminated by photons of energy equal to or greater than the semiconductor band gap, free-electron-hole pairs are generated in their active layers. This process increases the concentration of the minority carriers, for example, the holes in an n-type channel. This increase in hole concentration, Δp is proportional to αd, if $\alpha d < 1$. Mathematically, Δp is expressed as [49]:

$$\Delta p = \frac{\tau}{d}\left[\frac{P_{opt}\lambda}{hc}\right]\left[1 - \exp(-\alpha d)\right]. \quad (4.76)$$

where

h = Planck's constant,
P_{opt} = incident optical power per unit area,
λ = wavelength of the incident light,
α = optical absorption coefficient of the semiconductor,
d = thickness of the active layer,
τ = minority carrier lifetime,
c = speed of light in vacuum.

The light-induced voltage, V_{lit}, at the Schottky gate as a consequence of the excess hole concentration, is expressed as [49]:

$$V_{lit} = \frac{kT}{q}\ln\left[\frac{p + \Delta p}{p}\right] \quad (4.77)$$

where

k = Boltzmann's constant,
T = temperature in kelvins,
q = electron charge.

Thus, illuminating the gate region has the same effect as forward biasing the gate. The variable p in (4.77) is the equilibrium minority carrier concentration in the active layer and is given by

$$p = \frac{n_i^2}{n} \quad (4.78)$$

where n_i is the intrinsic carrier concentration, and n is the carrier concentration.

The optical performance of these devices is also characterized by the responsivity R, which is expressed as

$$R = \frac{I_p}{P_{opt}} \quad (4.79)$$

where I_p is the drain current when the incident optical power is P_{opt}.

The quantum efficiency η is the number of electron-hole pairs generated per incident photon of energy hc/λ and is related to R as follows:

$$\eta = \frac{hcR}{\lambda q} \qquad (4.80)$$

The light-induced voltage is experimentally determined by plotting the measured gate current I_g as a function of the reverse biased gate-to-source voltage V_{gs}, with the drain open circuited, and extrapolating the curve until it intersects the abscissa. The intersection point is V_{lit}. Figure 4.24 shows that V_{lit} is 0.57 V for the AlGaAs/GaAs HEMT (MPD-H503, Gould) when the optical illumination ($\lambda = 0.83$ μm) is approximately 1.7 mW [49]. The light-induced voltage is independent of the gate-to-source and gate-to-drain spacings, and is a function of the device material characteristics for a fixed incident optical power.

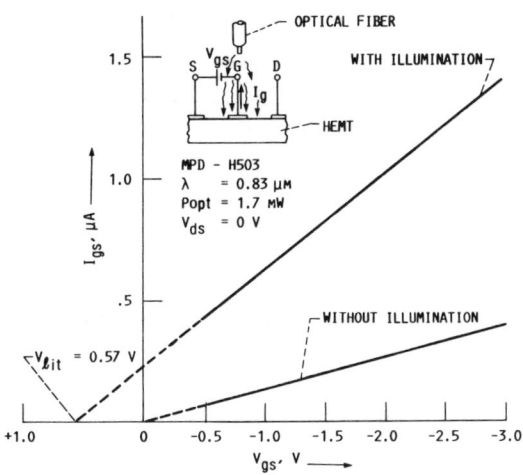

Figure 4.24 Measured I_g versus V_{gs} for AlGaAs-GaAs HEMT from which V_{lit} is determined.

The measured drain-to-source current, I_{ds}, as a function of the drain-to-source voltage, V_{ds} with and without optical illumination for a AlGaAs-GaAs HEMT (MPD-H503, Gould), is presented in Figure 4.25. This figure shows that the increase in the drain current at a fixed illumination for discrete values of V_{gs} is approximately 5 mA [49].

The responsivity of the HEMT is 3.53 A/W (when $V_{gs} = -0.5$ V, $V_{ds} = 3.0$ V, $I_{ds} = 6$ mA and $P_{opt} = 1.7$ mW), which, when compared with a calibrated AlGaAs-GaAs PIN photodiode, results in an external quantum efficiency of greater than 500% [49]. The measured dc transconductance, g_m is observed to be insensitive to optical illumination. The maximum change observed with illumination is less than 2 mS [49].

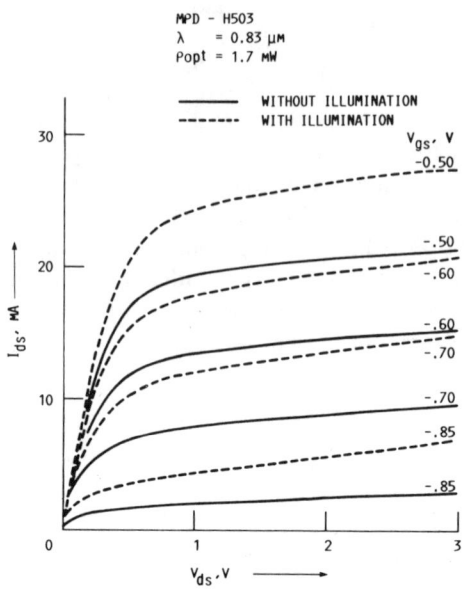

Figure 4.25 Measured I_{ds} versus V_{ds} for AlGaAs/GaAs HEMT with and without optical illumination.

4.7.3 FET Microwave Characteristics Under Optical Illumination

The small-signal, lumped-element equivalent circuit model [57] for the AlGaAs-GaAs HEMT, operated in the saturated current region and in the common source configuration, is shown in Figure 4.26(a). The circuit elements R_s, R_d, R_g, C_{gs}, C_{dg}, and C_{ds} have their physical origin within the device, and respectively represent the source-to-channel resistance, drain-to-channel resistance, gate metal resistance, gate-to-channel depletion capacitance, drain-to-gate feedback capacitance, and drain-to-source capacitance. The circuit elements R_i, Y'_m, and R'_0 are associated with the Schottky-barrier region and represent the charging channel resistance to C_{gs}, the transadmittance of magnitude g_{m0}, and phase delay τ (reflecting the carrier transit time in the channel section), and the channel resistance, respectively. Minasian [57] has shown that for a practical device $R_s/R'_0 \ll 1$ and Re $Y'_{12} \ll$ Im Y'_{12}. Hence, without loss of accuracy, the model of Figure 4.26(a) can be transformed into a π equivalent model and is shown in Figure 4.26(b). This model has been independently verified and found to be accurate up to 18 GHz [58]. The advantage of this model is that it is simple and conveniently related to the real and imaginary parts of the admittance parameters. The expressions are [57]:

$$C_d = -[\text{Im } Y_{12}]/\omega$$

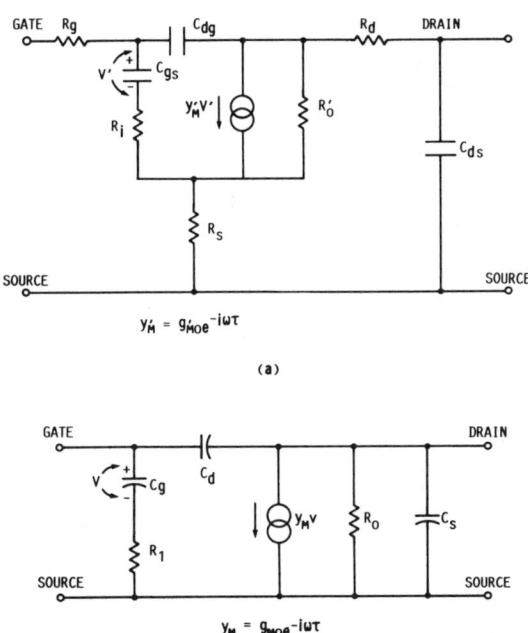

Figure 4.26 (a) A lumped-element equivalent circuit model for AlGaAs-GaAs HEMT; (b) a π equivalent circuit model.

$$C_s = [\text{Im } Y_{22}]/\omega - C_d$$
$$R_0 = 1/\text{Re } Y_{22}$$
$$g_{m0} = \text{Re } Y_{21}|\omega \to 0$$
$$C_g = [\text{Im } Y_{11}]/\omega - C_d$$
$$R_1 = [\text{Re } Y_{11}] (\omega C_g)^2$$
$$\tau = (-[\text{Im } Y_{21}]/\omega - g_{m0} R_1 C_g - C_d)/g_{m0} \qquad (4.81)$$

By substituting the measured admittance parameters [49] of AlGaAs/GaAs HEMT (MPD-H503, Gould) in the above expressions, the changes in the equivalent circuit element values are computed and illustrated as a function of the gate bias voltage in Figure 4.27. These figures show that the gate and source capacitances increase, while the drain-to-gate feedback capacitance decreases under optical illumination [49]. The gate charging resistance and the channel resistance both decrease under optical illumination [49].

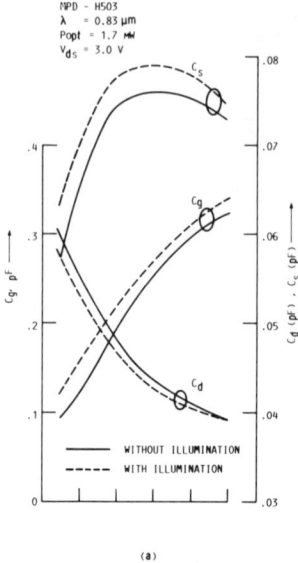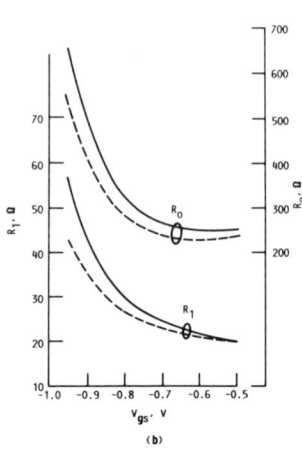

Figure 4.27 De-embedded lumped elements for AlGaAs-GaAs HEMT with and without optical illumination: (a) gate, source, and drain-to-gate feedback capacitances; (b) charging resistance and channel resistance.

Alternatively, we could also use the de-embedded scattering parameters to derive a small-signal, lumped-element equivalent circuit model [59] of the device, with and without optical illumination. This can be done by using EEsof's Touchstone software package [60]. With this technique, the computed maximum available gain and current gain ($|h_{21}|$) from the de-embedded scattering parameters of a GaAs MESFET (DXL 0503A, Gould) show that MAG is unaffected by optical illumination (λ = 0.83 μm). However, $|h_{21}|$ increases by a few dB under optical illumination of 1.5 mW [61]. The maximum frequency of oscillation (F_{max}) and the unity current gain cut-off frequency (f_t) obtained by extrapolating the MAG and $|h_{21}|$ curves, respectively, show that the F_{max} is insensitive to optical illumination, but f_t increases by a few GHz [61].

4.7.4 FET Noise Characteristics Under Optical Illumination

The noise spectrum of the AlGaAs/GaAs HEMT device without optical illumination is dominated by $1/f$ noise at frequencies below a few tens of MHz; at high frequencies, the channel conductance contributes to the noise, which is expressed as [62]:

$$i_n^2 = 4kTg_m G\Delta f \qquad (4.82)$$

where g_m is the transconductance of the HEMT, Δf is the bandwidth, and G is a quantity approximately equal to 1.1 [63]. The computed noise current at room temperature and with a g_m of 40 is found to be 27 pA/$\sqrt{\text{Hz}}$.

REFERENCES

1. Dolfi, D.W., M. Nazarathy, and R.L. Jungerman, "40 GHz Electro-Optic Modulator with 7.5 V Drive Voltage," *Electronics Letters*, Vol. 24, No. 9, April 1988, pp. 528–529.
2. Bowers, J.E., "Millimeter-Wave Response of InGaAsP Lasers," *Electronics Letters*, Vol. 21, 1985, p. 1195.
3. Keiser, G., *Optical Fiber Communications*, New York, McGraw-Hill, 1983, Chapter 6.
4. Sze, S.M., *Physics of Semiconductor Devices*, 2nd Ed., New York, John Wiley and Sons, 1981, Chapter 13, Section 13.2, Section 13.3.2.
5. Schlafer, J., C.B. Su, W. Powazinik, and R.B. Lauer, "20 GHz Bandwidth InGaAs Photodetector for Long-Wavelength Microwave Optical Links," *Electronics Letters*, Vol. 21, No. 11, May 1985, pp. 469–471.
6. Bowers, J.E., C.A. Burrus, and F. Mitschke, "Millimeter Waveguide Mounted InGaAs Photodectors," *Electronics Letters*, Vol. 22, No. 12, June 1986, pp. 633–635.
7. Lee, T.P., C.A. Burrus, A.G. Dentai, and K. Ogawa, "Small Area InGaAs/InP p-i-n Photodiodes: Fabrication, Characteristics and Performance of Devices in 274 Mb/s and 45 Mb/s Lightwave Receivers at 1.31 μm Wavelength," *Electronics Letters*, Vol. 16, No. 4, February 1980, pp. 155–156.
8. Lee, T.P., Photodetectors, Chapter 5 in Daly, J.C., ed., *Fiber Optics*, CRC Press, Florida, 1984, pp. 123–150.
9. EEsof Touchstone, EEsof, Inc., 1987.
10. Tucker, R.S., A.J. Taylor, C.A. Burrus, G. Eisenstein, and J.M. Wiesenfeld, "Coaxially Mounted 67 GHz Bandwidth InGaAs PIN Photodiode," *Electronics Letters*, Vol. 22, No. 17, August 1986, pp. 917–918.
11. Matthews, J.W., and A.E. Blakeslee, "Defects in Epitaxial Multilayers, *J. Crystal Growth*, Vol. 27, 1974, pp. 118–125.
12. Olsen, G.H., "Low-Leakage, High-Efficiency, Reliable VPE InGaAs 1.0–1.7 μm Photodiodes," *IEEE Electron Devices Letters*, Vol. EDL-2, No. 9, September 1981, pp. 217–219.
13. Robertson, M.J., S. Ritchie, S.K. Sargood, A.W. Nelson, L. Davis, R.H. Walling, C.P. Skrimshire, and R.R. Sutherland, "Highly Reliable Planar GaInAs/InP Photodiodes with High Yield Made by Atmospheric Pressure MOVPE," *Electronics Letters*, Vol. 24, No. 5, March 1988, pp. 252–254.
14. Bauer, J.G., and R. Trommer, "Long-Term Operation of Planar InGaAs/InP p-i-n Photodiodes," *IEEE Trans. Electron Devices*, Vol. ED-35, No. 12, December 1988, pp. 2349–2353.
15. Beneking, H., "Gain and Bandwidth of Fast Near-Infrared Photodetectors: A Comparison of Diodes, Phototransistors, and Photoconductive Devices," *IEEE Trans. Electron Devices*, Vol. ED-29, No. 9, September 1982, pp. 1420–1431.
16. Slayman, C.W., and L. Figueroa, "Frequency and Pulse Response of a Novel High Speed Interdigital Surface Photoconductor (IDPC)," *IEEE Electron Devices Letters*, Vol. EDL-2, No. 5, May 1981, pp. 112–114.
17. Lee, W.S., G.R. Adams, J. Mun, and J. Smith, "Monolithic GaAs Photoreceiver for High-Speed Signal Processing Applications," *Electronics Letters*, Vol. 22, No. 3, January 1986, pp. 147–148.

18. Jackson, D.J., and D.L. Persechini, "Monolithically Integrable High Speed Photodetectors," *High Frequency Optical Communications,* SPIE, Vol. 716, 1986, pp. 104–108.
19. Lam, D.K.W., R.I. MacDonald, J.P. Noad, B.A. Syrett, "Surface-Depleted Photoconductors," *IEEE Trans. Electron Devices,* Vol. ED-34, No. 5, May 1987, pp. 1057–1059.
20. Ito, M., and O. Wada, "Low Dark Current GaAs Metal-Semiconductor-Metal (MSM) Photodiodes Using WSi_x Contacts," *IEEE J. Quantum Electronics,* Vol. QE-22, No. 7, July 1986, pp. 1073–1077.
21. Roth, W., H. Schumacher, J. Kluge, H.J. Geelen, and H. Beneking, "The DSI Diode—A Fast Large-Area Optoelectronic Detector," *IEEE Trans. Electron Devices,* Vol. ED-32, No. 6, June 1985, pp. 1034–1036.
22. Wojtczuk, S.J., and J.M. Ballantyne, "Impedance Properties and Broad-Band Operation of GaAs Photoconductive Detectors," *IEEE J. Lightwave Technol.,* Vol. LT-5, No. 3, March 1987, pp. 320–324.
23. Golio, J.M., and R.J. Trew, "Optimum Semiconductors for High-Frequency and Low-Noise MESFET Applications," *IEEE Trans. Electron Devices,* Vol. ED-30, No. 10, October 1983, pp. 1411–1413.
24. Kajiyama, K., Y. Mizushima, and S. Sakata, "Schottky-Barrier Height of n-$In_xGa_{1-x}As$ Diodes," *Applied Physics Letters,* Vol. 23, No. 8, October 1973, pp. 458–459.
25. Chen, C.Y., B.L. Kasper, H.M. Cox, and J.K. Plourde, "2-Gb/s Sensitivity of a $Ga_{.47}In_{.53}As$ Photoconductive Detector/GaAs Field-Effect Transistor Hybrid Photoreceiver," *Applied Physics Letters,* Vol. 46, No. 4, February 1985, pp. 379–381.
26. Rao, M.V., P.K. Bhattacharya, and C.Y. Chen, "Low-Noise $In_{.53}Ga_{.47}As$:Fe Photoconductive Detectors for Optical Communication," *IEEE Trans. Electron Devices,* Vol. ED-33, No. 1, January 1986, pp. 67–71.
27. Kim, J.H., S.S. Li, and J.J. Pan, "Picosecond Response High-Speed GaAs Schottky-barrier Photodetector for Microwave Optical Fiber Links," *High Frequency Optical Communications,* SPIE, Vol. 716, 1986, pp. 96–103.
28. Parker, D.G., P.G. Say, A.M. Hansom, and W. Sibbett, "110 GHz High-Efficiency Photodiodes Fabricated from Indium Tin Oxide/GaAs," *Electronics Letters,* Vol. 23, No. 10, May 1987, pp. 527–528.
29. Parker, D.G., "Use of Transparent Indium Tin Oxide to Form a Highly Efficient 20 GHz Schottky-barrier Photodiode," *Electronics Letters,* Vol. 21, No. 18, August 1985, p. 778.
30. Selders, J., N. Emeis, and H. Beneking, "Schottky-Barrier on p-type GaInAs," *IEEE Trans Electron Devices,* Vol. ED-32, No. 3, March 1985, pp. 605–609.
31. Emeis, N., H. Schumacher, and H. Beneking, "High-Speed GaInAs Schottky Photodetector," *Electronics Letters,* Vol. 21, No. 5, February 1985, pp. 180–181.
32. Wang, S.Y., D.M. Bloom, and D.M. Collins, "GaAs Schottky Photodiode with 3-dB Bandwidth of 20 GHz," *IEEE Int. Electron Devices Meeting Tech. Digest,* 1982, pp. 521–524.
33. Wang, S.Y., and D.M. Bloom, "100 GHz Bandwidth Planar GaAs Schottky Photodiode," *Electronics Letters,* Vol. 19, No. 14, July 1983, pp. 554–555.
34. Parker, D.G., and P.G. Say, "Indium Tin Oxide/GaAs Photodiodes for Millimetric-Wave Applications," *Electronics Letters,* Vol. 22, No. 23, November 1986, pp. 1266–1267.
35. Blauvelt, H., G. Thurmond, J. Parsons, D. Lewis, and H. Yen, "Fabrication and Characterization of GaAs Schottky-Barrier Photodetectors for Microwave Fiber Optic Links," *Applied Physics Letters,* Vol. 45, No. 3, August 1984, pp. 195–196.
36. Liao, S.Y., *Microwave Solid-State Devices,* Prentice-Hall, Englewood Cliffs, NJ, 1985, p. 15.
37. Willardson, R.K., and A.C. Beer (eds.), *Semiconductors and Semimetals, Vol. 14, Lasers, Junctions, Transport,* New York, Academic Press, 1979, p. 125.
38. Campbell, J.C., A.G. Dentai, W.S. Holden, and B.L. Kasper, "High-Performance Avalanche Photodiode with Separate Absorption Grading and Multiplication Regions," *Electronics Letters,* Vol. 19, No. 20, September 1983, pp. 818–820.

39. Campbell, J.C., B.C. Johnson, G.J. Qua, and W.T. Tsang, "Frequency Response of InP/InGaAsP/InGaAs Avalanche Photodiodes," *IEEE J. Lightwave Technol.*, Vol. LT-7, No. 5, May 1989, pp. 778–784.
40. Holden, W.S., J.C. Campbell, J.F. Ferguson, A.G. Dentai, and Y.K. Jhee, "Improved Frequency Response of InP/InGaAsP/InGaAs Avalanche Photodiodes with Separate Absorption, Grading and Multiplication Regions," *Electronics Letters*, Vol. 21, No. 20, September 1985, pp. 886–887.
41. Forrest, S.R., "Optical Detectors for Long-Wavelength Lightwave Communication," *Components for Fiber Optic Applications*, SPIE, Vol. 722, 1986, pp. 188–191.
42. J.C. Gammel and J.M. Ballantyne, "The OPFET: A New High Speed Optical Detector," *IEEE Int. Electron Devices Meeting Tech. Digest*, December 1978, pp. 120–123.
43. Sugeta, T., and Y. Mizushima, "High Speed Photo Response Mechanism of a GaAs-MESFET," *Japan J. Applied Physics*, Vol. 19, January 1980, pp. L27–L29.
44. Mizuno, H., "Microwave Characteristics of an Optically Controlled GaAs MESFET," *IEEE Trans. Microwave Theory Tech.*, Vol. MTT-31, July 1983, pp. 596–600.
45. DeSalles, A.A.A., "Optical Control of GaAs MESFET's," *IEEE Trans. Microwave Theory Tech.*, Vol. MTT-31, October 1983, pp. 812–820.
46. Gautier, J.L., D. Pasquet, and P. Pouvil, "Optical Effects on the Static and Dynamic Characteristics of a GaAs MESFET," *IEEE Trans. Microwave Theory Tech.*, Vol. MTT-33, September 1985, pp. 819–822.
47. Chen, C.Y., A.Y. Cho, C.G. Bethea, P.A. Garbinski, Y.M. Pang, B.F. Levine, and K. Ogawa, "Ultrahigh Speed Modulation-Doped Heterostructure Field-Effect Photodetectors," *Applied Physics Letters*, Vol. 42, June 1983, pp. 1040–1042.
48. Umeda, T., Y. Cho, and A. Shibatomi, "Picosecond HEMT Photodetector," *Japan J. Applied Physics*, Vol. 25, October 1986, pp. L801–L803.
49. Simons, R.N., "Microwave Performance of an Optically Controlled AlGaAs/GaAs High Electron Mobility Transistor and GaAs MESFET," *IEEE Trans. Microwave Theory Tech.*, Vol. MTT-35, No. 12, December 1987, pp. 1444–1455.
50. Fukui, H., "Optimal Noise Figure of Microwave GaAs MESFET's," *IEEE Trans. Electron Devices*, Vol. ED-26, July 1979, pp. 1032–1037.
51. Chaim, N.B., S. Margalit, A. Yariv and I. Ury, "GaAs Integrated Optoelectronics," *IEEE Trans. Electron Devices*, Vol. ED-29, No. 9, September 1982, pp. 1372–1381.
52. Swanson, A., J. Herb, and M. Yung, "First Commercial HEMT Challenges GaAs FET's," *Microwaves & RF*, Vol. 24, November 1985, pp. 107–118.
53. Nguyen, L.D., D.C. Radulescu, P.J. Tasker, W.J. Schaff, and L.F. Eastman, "0.2-μm Gate-Length Atomic-Planar Doped Pseudomorphic $Al_{0.3}Ga_{0.7}As/In_{0.25}Ga_{0.75}As$ MODFET's with f_t over 120 GHz," *IEEE Electron Devices Letters*, Vol. EDL-9, No. 8, August 1988, pp. 374–376.
54. Mishra, U.K., A.S. Brown, S.E. Rosenbaum, C.E. Hooper, M.W. Pierce, M.J. Delaney, S. Vaughn, and K. White, "Microwave Performance of AlInAs-GaInAs HEMT's with 0.2- and 0.1-μm Gate Length," *IEEE Electron Device Letters*, Vol. EDL-9, No. 12, December 1988, pp. 647–649.
55. Ketterson, A.A., J. Laskar, T.L. Brock, I. Adesida, J. Kolodzey, O.A. Aina, and H. Hier, "Dependence of Current-Gain Cutoff Frequency on Gate Length in Submicron GaInAs/AlInAs MODFET's," *Electronics Letters*, Vol. ED-25, No. 7, March 1989, pp. 440–442.
56. Lepore, A., M. Levy, H. Lee, E. Kohn, D. Radulescu, R. Tiberio, P. Tasker, and L. F. Eastman, "Fabrication and Performance of 0.1-μm Gate-Length AlGaAs/GaAs HEMT's with Unity Current Gain Cutoff Frequency in Excess of 110 GHz," *46th Annual Device Research Conf. Abstract, IEEE Trans. Electron Devices*, Vol. ED-35, No. 12, December 1988, pp. 2441–2442.
57. Minasian, R.A., "Simplified GaAs MESFET Model to 10 GHz," *Electronics Letters*, Vol. 13, September 1977, pp. 549–551.

58. Reeder, T.M., and W.B. Wylie, "CAD/CAM for GaAs IC's," *Proc. Microwave Systems Applied Technology Conf.*, March 1983, pp. 461–474.
59. Simons, R.N., and R.R. Romanofsky, "Microwave Characterization and Modeling of GaAs/AlGaAs Heterojunction Bipolar Transistors," *EEsof User's Group Meeting*, Las Vegas, NV, June 9, 1987; see also *NASA Tech. Memo. 100150*.
60. *EEsof Touchstone User's Manual*, EEsof, 1987.
61. Simons, R.N., *Optoelectronic Gain Control of a Microwave Single Stage GaAs MESFET Amplifer, NASA Contractor Report 182201*, September 1988.
62. Van der Ziel, A., *Noise, Sources, Characterization, and Measurements*, Englewood Cliffs, NJ, Prentice-Hall, 1970.
63. Baechtold, "Noise Behavior of GaAs Field Effect Transistors with Short Gate Lengths," *IEEE Trans. Electron Devices*, Vol. ED-19, May 1972, pp. 674–680.
64. Zeghbroeck, B.J.V., W. Patrick, J.M. Halbout, and P. Vettiger, "105-GHz Bandwidth Metal-Semiconductor-Metal Photodiode," *IEEE Electron Devices Letters*, Vol. EDL-9, No. 10, October 1988, pp. 527–529.

Chapter 5
Microwave Fiber Optic Links

5.1 INTRODUCTION

Microwave fiber optic communication links have several advantages when compared to conventional coaxial and waveguide links. The obvious advantages are the greatly reduced size and weight. These advantages can be appreciated by comparing the size and weight of a rectangular waveguide and also a semirigid coaxial cable with an optical fiber cable. Table 5.1 presents this comparison.

Another feature that is derived from Table 5.1 is that as the number of transmission paths increases, the weight saving obtained with optical fiber cable increases, dramatically. As an example, a 100 feet long transmission system containing 10 transmission paths would weigh approximately 32.5 lbs if implemented in coaxial cable. If the transmission system were to be implemented in optical fiber cable it would weigh only 5.5 lbs, resulting in 83% savings in weight. This weight saving feature is especially significant in space applications, such as a phased array harness.

The third advantage of an optical fiber is the attenuation, which remains low and constant over the entire microwave frequency range. However, the attenuation in coaxial cable and rectangular waveguide tends to increase with frequency due to the frequency dependence of the skin effect. Figure 5.1 compares the attenuation of the various transmission lines over the microwave frequency range. Finally, fiber optic links are impervious to *electromagnetic interference* (EMI), *radio frequency interference* (RFI), *electromagnetic pulse* (EMP), and snooping. Applications of fiber optic links include radar, communication, signal processing, and remote sensing.

In Section 5.2 the definitions of various losses occurring in a typical fiber optic link are explained. In general, microwave fiber optic communication links are of two types and they are known by the technique that is employed to intensity modulate the laser light. In a direct modulation link, the laser diode is intensity-modulated by imposing the microwave signal on the laser dc bias current. Conversely, in an external modulation type of link, the laser operates in a CW mode and an external modulator imposes the microwave signal on the optical carrier. Section 5.2 presents

Table 5.1
Characteristics of Various Microwave Transmission Media Over a Distance of 100 Feet

Transmission Media	Outside Dimensions (inches)	Weight per 100 ft (lbs)	Attenuation	Reference
Fiber Optic Cable 50 μm/125 μm (Part #227101)	0.150 diameter	0.55	4.0 dB/km at 1.3 μm	[1]
Aluminium Rectangular Waveguide (WR-90)	1.0 × 0.5	16.38	5.49 dB/100 ft at 8.2 GHz, and 3.83 dB/100 ft at 12.4 GHz	[2,3]
Semirigid Coaxial Cable (RG-402U) (part #BP50141)	0.141 diameter	3.25	44.5 dB/100 ft at 10 GHz, and 62.0 dB/100 ft at 18 GHz	[4]

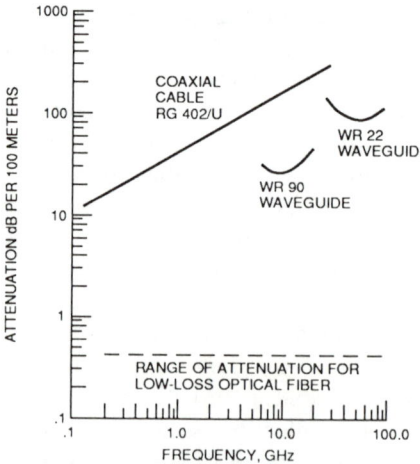

Figure 5.1 Microwave transmission media attenuation *versus* frequency.

analytical expressions to compute the insertion loss of these links. The shot noise processes associated with carrier injection and recombination inside the laser diode active layer cause fluctuations in the output-light intensity. The intensity fluctuations result in intensity noise which degrades the signal-to-noise ratio. Hence, Section 5.3

explains the relative intensity noise of a typical semiconductor laser diode. Section 5.4 presents detailed analysis and computed results on the link noise figure. The link bandwidth, harmonic and intermodulation distortion, and, spurious-free dynamic range and maximum signal-to-noise ratio are discussed in Sections 5.5 through 5.7, respectively. Single-mode optical fiber dispersion is briefly explained in Section 5.8, while Section 5.9 discusses the optical losses in a microwave fiber optic link. The purpose of Section 5.9 is to present typical data on losses occurring in optical interconnecting components which are subsequently used in link design. (Radiation hardness and reliability of laser diodes, photodiodes and optical fibers are presented in the chapters on laser diodes, photodiodes, and optoelectronic switch matrix, and therefore will not be discussed here.) The performance of typical directly-modulated and also externally-modulated RF fiber optic links are compared in Section 5.10. Finally, Section 5.11 presents fiber optic link design examples.

5.2 INSERTION LOSS OF A MICROWAVE FIBER OPTIC LINK

A simple fiber optic link consisting of a semiconductor laser diode, a length of an optical fiber, and a photodiode is shown in Figure 5.2. In this figure, several sources of optical and optoelectronic loss along the link are indicated. The bias current, I_L, flowing into the laser diode is converted into optical power and is expressed in terms of the conversion efficiency, η_L, of the laser. This quantity is determined from the slope of the optical output power *versus* bias current curve of the laser diode. Similarly, the optical power incident on the active area of the photodiode is converted into current and is expressed in terms of the conversion efficiency, η_P, of the photodiode. This quantity is determined from the slope of the bias current *versus* optical power curve of the photodiode. A fraction of the power emitted by the laser diode is lost due to imperfect coupling between the laser diode and the optical fiber. The diode-to-fiber coupling efficiency is denoted as η_{LF}. Similarly, the fiber-to-photodiode coupling efficiency is denoted as η_{FP}. The optical fiber loss η_F is dependent

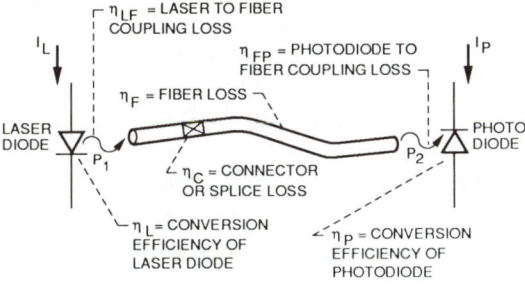

Figure 5.2 A typical fiber optic link setup showing the various optical as well as optoelectronic losses.

on the wavelength of operation and is much smaller than the coupling loss. The connector or splice loss is denoted as η_C. By considering all of the above factors, a transfer coefficient η is defined between the dc current, I_L, flowing into the laser diode and the resulting current, I_P, generated at the photodiode. That is,

$$I_P = \eta I_L \tag{5.1}$$

The current transfer coefficient, η, from the laser to the detector is the product of all the optical and optoelectronic losses in the link and is expressed as

$$\eta = \eta_L \, \eta_{LF} \, \eta_F \, \eta_C \, \eta_{FP} \, \eta_P \tag{5.2}$$

Equation (5.1) can also be written in terms of an electrical power transfer function as follows:

$$P_P = \xi^2 P_L \tag{5.3}$$

Where P_L and P_P are the electrical power supplied and delivered to the laser diode and by the photodiode respectively, and ξ^2 is the electrical power transfer coefficient from the laser to the photodiode. Also, ξ is expressed as

$$\xi = \frac{\eta_L}{\sqrt{R_L}} \xi_F \, \eta_P \sqrt{R_P} \tag{5.4}$$

where

R_L = laser diode incremental drive impedance about its point of bias;
R_P = photodiode terminating load impedance;
ξ_F = optical fiber power transfer efficiency, which includes optical coupling losses at both the laser and photodiode and the connector or splice loss;
= $\eta_{LF} \, \eta_F \, \eta_C \, \eta_{FP}$.

The optical insertion loss of the link can now be defined as

$$\text{dB (optical)} = 10 \log(P_2/P_1) \tag{5.5}$$

where P_1 and P_2 are the optical power emitted and delivered by the laser diode and to the photodiode, respectively. Because the optical power output from the laser is proportional to the bias current and the photocurrent proportional to the incident optical power, (5.5) is equal to

$$10 \log(P_2/P_1) = 10 \log(I_P/I_L) \tag{5.6}$$

Insofar as the electrical power delivered to the load is proportional to the square of the current, the electrical insertion loss of the link is

$$\text{dB (electrical)} = 10 \log(I_P/I_L)^2$$
$$= 20 \log(I_P/I_L)$$
$$= 2 \times \text{dB (optical)} \qquad (5.7)$$

Thus, the optical and optoelectronic losses contribute as their square to the overall RF link loss.

The fiber optic link shown in Figure 5.2 is a simplified setup of a link used for microwave signal transmission. Figures 5.3 and 5.4 schematically illustrate the two principal approaches that are used for microwave signal transmission. These links are designated by the technique that is used to modulate the optical carrier. The optical carrier can be modulated either by directly modulating the laser diode or by the use of an external modulator.

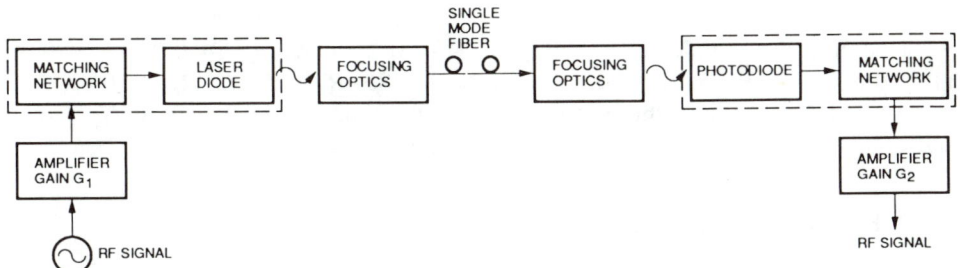

Figure 5.3 Schematic of a directly modulated analog fiber optic links for microwave transmission.

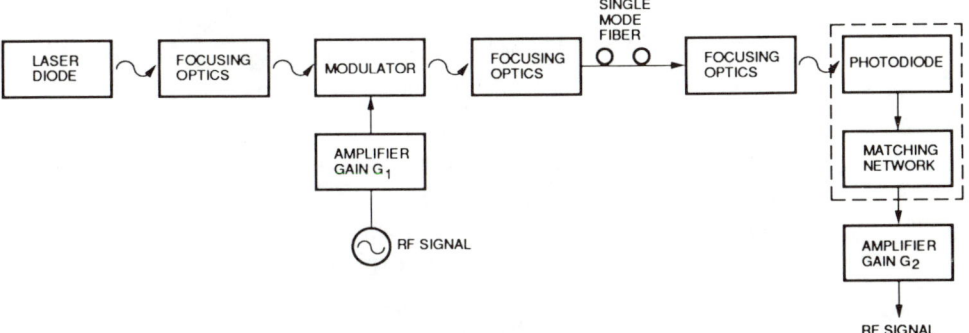

Figure 5.4 Schematic of an externally modulated analog fiber optic links for microwave transmission.

5.2.1 Directly Modulated Link

For a directly modulated link the RF insertion loss is expressed as [5]:

$$\text{Insertion Loss (dB)} = 10 \log(\text{power out/power in}) \text{ dB}$$

$$= 10 \log\left[\frac{\rho_L K \rho_P R_2 G_{1DM} G_{2DM}}{R_1}\right] \text{ dB} \quad (5.8)$$

where

$K = (\eta_L \eta_{LF} \eta_C \eta_F \eta_{FP} \eta_P)^2$
ρ_L = laser diode matching circuit loss,
ρ_P = photodiode matching circuit loss,
R_1 = source impedance,
R_2 = load impedance,
G_{1DM} = gain of the input amplifier,
G_{2DM} = gain of the output amplifier.

As a numerical example, if $\rho_L = 0.87$, $\eta_L = 0.2$ W/A, $\eta_{LF} = 0.2$, $\eta_C = 0.96$, $\eta_F = 0.8$, $\eta_{FP} = 0.85$, $\eta_P = 0.8$ A/W, $\rho_P = 0.85$, $R_1 = 50 \, \Omega$, $R_2 = 50 \, \Omega$, $G_{1DM} = 1.0$, and $G_{2DM} = 1.0$, the insertion loss of the link is -34.91 dB. If $G_{1DM} = 10$ and $G_{2DM} = 100$, the link insertion loss reduces to -4.91 dB.

5.2.2 Externally Modulated Link

For an externally modulated link, the insertion loss is expressed as [5]:

Insertion Loss (dB) = 10 log(power output/power input) dB

$$= 10 \log\left[G_{1EOM} G_{2EOM} (1 - |\Gamma|^2) R_1 R_2 \left[\frac{P_1 \eta_C \eta_M \eta_F \eta_P}{2V_\pi}\right]^2\right] \text{ dB} \quad (5.9)$$

where

Γ = reflection coefficient due to the transition between the 50 Ω line and the modulator input impedance,
η_M = optical loss through the modulator,
V_π = voltage for 100% modulation as explained in Section 3.7.1 of Chapter 3 on electro-optic modulators,
G_{1EOM} = gain of the input amplifier,
G_{2EOM} = gain of the output amplifier.

As a numerical example, if $G_{1EOM} = 1.0$, $G_{2EOM} = 1.0$, $R_1 = 50\ \Omega$, $R_2 = 50\ \Omega$, $V_\pi = 8.0$ V (rms), $P_1 = 1.7$ mW, $\eta_M = 0.42$, $\eta_C = 0.409$, $\eta_F = 0.855$, and $\eta_P = 0.75$, the insertion loss of the link is -64.65 dB.

5.3 RELATIVE INTENSITY NOISE OF A SEMICONDUCTOR LASER DIODE

The measured spectrum of conventional double-heterostructure lasers as discussed in Section 2.5.2, indicate the presence of several longitudinal modes. The output intensity of each of these longitudinal modes is observed to change abruptly in a random fashion, although the total output light intensity of the laser appears to be stabilized. This phenomenon results from fluctuations in the partition of the total laser output among different longitudinal modes [6].

The laser optical power fluctuations can be analytically described with the help of the laser rate equations such as those presented in Section 2.7. To these basic equations, additional terms representing noise generators are introduced to describe the shot noise associated with carrier injection and recombination in the laser active medium. These equations can then be numerically solved to yield the fluctuations in intensity of the individual modes. The fluctuations in the output intensity give rise to a *relative intensity noise* (RIN) spectrum in the power output from the laser. Mathematically RIN is expressed as [7]:

$$\text{RIN}(f) = \frac{\langle \Delta P^2(f) \rangle}{P_L^2}. \tag{5.10}$$

where P_L is the steady-state optical power output from the laser and $\langle \Delta P^2(f) \rangle$ is the spectral density of the square of the laser optical power fluctuation. The frequency f represents the low-frequency fluctuations of unmodulated laser diodes or the modulation frequency, in the case of intensity modulated laser diodes. As the solution of the rate equations does not yield closed-form expressions, the discussions presented here are qualitative.

The noise intensity of individual longitudinal modes has also been observed to be anticorrelated below the relaxation oscillation frequency, f_r of the laser [6]. This is because all the longitudinal modes are sustained by a common active medium which acts as an energy reservoir. Consequently, the RIN of all the longitudinal modes is typically about 40 dB lower than that of an individual longitudinal mode. Figure 5.5 schematically illustrates the typical variation of RIN spectrum of the dominant longitudinal mode as well as the sum total of all longitudinal modes.

Furthermore, as the bias current is increased above threshold, more numbers of lasing longitudinal modes appear in the output spectrum of the laser. Consequently, the RIN of all the modes is expected to decrease significantly as the bias

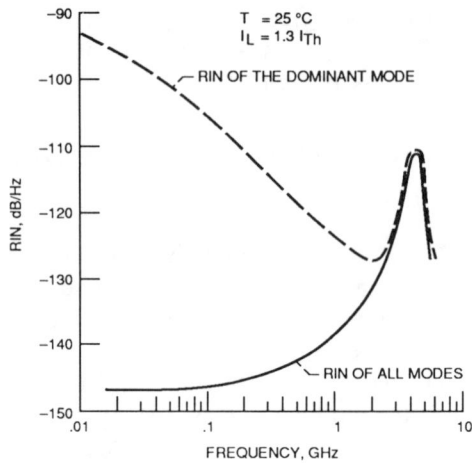

Figure 5.5 Typical variation of the relative intensity noise of the dominant mode as well as the sum total of all modes of a laser diode.

current is increased. Figure 5.6 illustrates the typical variation of RIN spectrum as a function of the bias current through the laser diode. We can infer from Figure 5.6 that the amplitude of the RIN spectrum has a broad and pronounced resonance peak at the relaxation oscillation frequency of the laser.

By employing a *buried heterostructure distributed feedback* (BH-DFB) laser diode similar to that described in Section 2.9.3, it is possible to stabilize the light intensity of the main mode. Consequently, the RIN is observed to be typically about 30 dB lower than conventional *buried heterostructure* (BH) laser diode for identical modulation depths, frequency and bias current [8]. Figure 5.7 compares the typical variation of RIN of the dominant longitudinal mode, with modulation depth of a conventional BH laser diode with that of a BH-DFB laser diode.

The noise spectra shown in Figure 5.6 represents the laser RIN under condi-

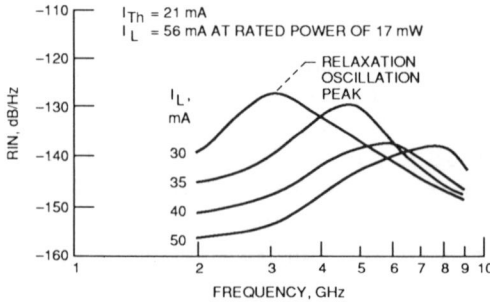

Figure 5.6 Typical variation of the relative intensity noise of a laser diode with bias current.

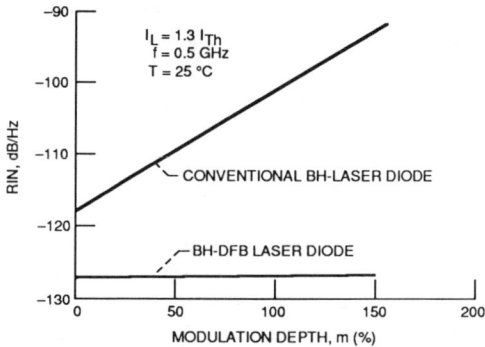

Figure 5.7 Typical variation of the relative intensity noise of the dominant mode as a function of the modulation depth for a conventional BH laser diode and BH-DFB laser diode.

tions of low optical feedback. In the presence of both near-end and far-end optical feedback, the RIN increases because the reflected power couples with the lasing modes, thereby causing their phases to vary. As an example, in the presence of excess optical reflections on the order of -10 dB from the near-end (few meters) of an optical fiber, the RIN increases by 10 dB. However, if this reflection occurred at the far-end (few kilometers) of an optical fiber the RIN would increase by about 5 dB [9]. Hence, optical isolators must be incorporated in the link to prevent the degradation of RIN due to reflection. The effect of RIN on the FM noise or phase noise of the system is explained in Section 7.7.2 of Chapter 7.

5.4 NOISE FIGURE OF A MICROWAVE FIBER OPTIC LINK

5.4.1 Directly Modulated Link

The analysis that is presented in this section is based on the work reported in [7]. The RIN as defined in equation 5.10 represents the total output noise from a laser diode. This includes the "shot noise" associated with carrier injection and recombination, input thermal noise from the RF signal source, and all other noise processes in the laser. The input thermal noise from the RF signal source is expressed as

$$N_{in} = KTB \tag{5.11}$$

where

$K = $ Boltzmann's constant,
$T = (273 + °C)$ is the absolute temperature, K;
$B = $ bandwidth.

Hence, the equivalent electrical noise power generated in the laser is the difference in the noise power of equations 5.10 and 5.11, which is

$$N_L(f) = \text{RIN}_{DM}(f)(I_{bias} - I_{th})^2 R_L B - KTB \qquad (5.12)$$

where

I_{bias} = bias current of the laser diode,
I_{th} = threshold current of the laser diode.

At the receiver end, the photodiode output electrical noise power, N_P, is the shot noise associated with the average photocurrent, I_P, which is expressed as

$$N_P = 2eI_p F_P B R_P \qquad (5.13)$$

where a factor $F_P \geq 1$ has been added to account for the excess noise in the case of avalanche photodiodes and e is the electronic charge. By using (5.1) through (5.4), (5.13) reduces to

$$N_P = 2e(I_{bias} - I_{th})\xi F_P B \sqrt{R_L R_P} \qquad (5.14)$$

The total electrical noise power at the output of the link, N_{out}, is the sum of the laser noise translated to the output, input RF signal source noise translated to the output and the photodiode noise. That is,

$$N_{out} = \xi^2 N_L + \xi^2 N_{in} + N_P \qquad (5.15)$$

By using (5.11), (5.12), and (5.14), and bearing in mind that the optical loss contributes as the square to the link loss:

$$N_{out} = \xi^2 [\text{RIN}_{DM}(f)(I_{bias} - I_{th})^2 R_L B - KTB]$$
$$+ \xi^2 KTB$$
$$+ 2e(I_{bias} - I_{th})\xi F_P B \sqrt{R_L R_P} \qquad (5.16)$$

The noise figure of the directly modulated link can be expressed as

$$F_{DM} = \frac{(S_{in}/N_{in})}{(S_{out}/N_{out})}$$
$$= \left(\frac{S_{in}}{S_{out}}\right)\left(\frac{N_{out}}{N_{in}}\right) \qquad (5.17)$$

where (S_{in}/N_{in}) and (S_{out}/N_{out}) are the signal-to-noise ratios at the input and the output of the link, respectively. Substituting from (5.3) and (5.11), then (5.16) yields

$$F_{DM} = \frac{\text{RIN}_{DM}(f)\,(I_{bias} - I_{th})^2\,R_L}{KT}$$

$$+ \frac{2e(I_{bias} - I_{th})\,F_P\,\sqrt{R_L R_P}}{\xi KT} \quad (5.18)$$

The noise figure given by the above expression can be attributed to the optoelectronic components in the link, which include the laser diode, photodiode, and optical fiber. In Figure 5.8, the noise figure F_{DM} is plotted as a function of the link transfer function ξ for different values of laser RIN. The values of the device parameters listed in Figure 5.8 are typical for a directly modulated link. Figure 5.8 shows that the noise figure attributable to the optoelectronic components is very large. The shaded portion in Figure 5.8 represents the range in which direct modulation links normally operate [7]. The dominant noise source in such links is the laser noise and therefore the link noise figure is directly proportional to the laser RIN, and relatively insensitive to the link transfer function ξ.

If amplifiers with noise figures F_1 and F_2, which are less than F_{DM}, are used before and after the link, respectively, the overall link noise figure improves. The

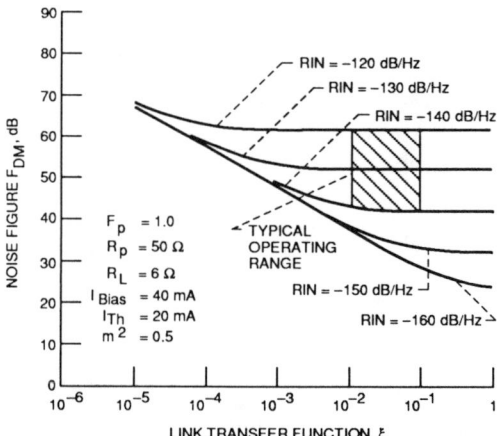

Figure 5.8 Computed noise figure attributable to the optoelectronic components of a directly modulated microwave link.
Source: Yen, H.W., C.M. Gee, and H. Blauvelt, "High-Speed Optical Modulation Techniques," *Optical Technology for Microwave Applications II*, SPIE, Vol. 545, 1985, pp. 2–9. Reprinted with permission.

overall link noise figure, F_{DML}, is determined from the well known Friis formula [10] for cascaded networks as follows:

$$F_{DML} = F_1 + \frac{F_{DM} - 1}{G_{1DM}} + \frac{F_2 - 1}{G_{1DM} \, \xi^2} \qquad (5.19)$$

where G_{1DM} is the gain of the input amplifier. To minimize the overall link noise figure F_{DML}, the RIN_{DM} should be minimized and the gain ξ^2 and G_{1DM} should be maximized. The RIN_{DM} can be minimized by selecting a laser diode with a low RIN. The gain ξ^2 can be maximized by lowering the optical and optoelectronic losses. However, the gain G_{1DM}, if increased arbitrarily, may lead to laser burnout and negative peak clipping. Hence, an optimum value for G_{1DM} is determined as explained below. Consider, the laser diode optical power output *versus* the bias current characteristic shown in Figure 5.9. This laser diode is assumed to be directly modulated by a sinusoidal RF signal about a bias current point I_{bias}. The modulation index m is defined as

$$m = \frac{\Delta I}{I_{\text{bias}} - I_{\text{th}}} \qquad (5.20)$$

where ΔI is the variation in the RF signal current about a bias point. From Figure 5.9 the laser diode maximum operating current I_{pk} can be expressed as

$$I_{\text{pk}} = I_{\text{th}} + 2 \, (I_{\text{bias}} - I_{\text{th}}). \qquad (5.21)$$

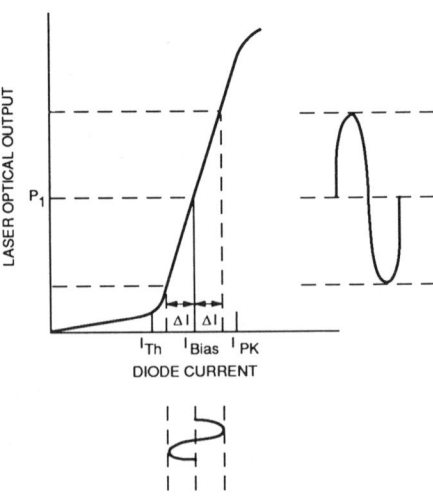

Figure 5.9 Laser diode optical output *versus* the bias current.

Substituting (5.21) in (5.20) yields

$$\Delta I = \frac{m(I_{pk} - I_{th})}{2} \quad (5.22)$$

Thus, the peak RF signal power, P_o, that can be dissipated in the laser diode incremental drive impedance, R_L, without waveform clipping is

$$P_o = \frac{\Delta I^2 R_L}{2} \quad (5.23)$$

Substituting (5.22) into (5.23) yields

$$P_o = \frac{m^2 (I_{pk} - I_{th})^2 R_L}{8} \quad (5.24)$$

Hence, if P_i is the input peak RF signal power to the amplifier from the signal source, the gain, G_{1DM}, of the amplifier should be set equal to

$$G_{1DM} = \frac{P_o}{P_i} \quad (5.25)$$

Substituting from (5.24) yields

$$G_{1DM} = \frac{m^2 (I_{pk} - I_{th})^2 R_L}{8 P_i} \quad (5.26)$$

Substituting (5.18) and (5.26) into (5.19) and using (5.21) yields

$$F_{DML} = F_1 - \left[\frac{8 P_i}{m^2 (I_{pk} - I_{th})^2 R_L}\right] + \left[\frac{2 P_i \, \text{RIN}_{DM}}{m^2 \, KT}\right]$$
$$+ \left[\frac{8e \sqrt{R_P/R_L} \, P_i \, F_P}{m^2 \, \xi KT (I_{pk} - I_{th})}\right] + \left[\frac{8 P_i (F_2 - 1)}{m^2 \, \xi^2 (I_{pk} - I_{th})^2 R_L}\right] \quad (5.27)$$

The directly modulated link noise figure is shown in Figure 5.10. These curves show that for large input signal levels the fiber optic link degrades the signal-to-noise ratio. However, for low input signal levels the link noise figure is given by the preamplifier noise figure F_1. The importance of using lasers with low RIN is also clearly indicated in Figure 5.10. Commercially available InGaAsP-InP distributed feedback laser diodes

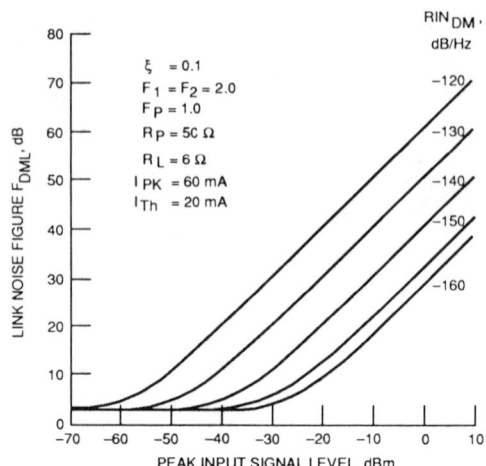

Figure 5.10 Computer overall link noise figure of a directly modulated microwave link.
Source: Yen, H.W., C.M. Gee, and H. Blauvelt, "High-Speed Optical Modulation Techniques," *Optical Technology for Microwave Applications II*, SPIE, Vol. 545, 1985, pp. 2–9. Reprinted with permission.

operating at 1.3 μm have typically a RIN of about −145 dB/Hz [11]. However, GaAlAs/GaAs buried heterostructure laser diodes operating at 0.84 μm wavelength have typically a RIN of about −120 dB/Hz [12].

5.4.2 Externally Modulated Link

The optical power output of the modulator is expressed as [7]:

$$P = P_{pk} \sin^2\left[\frac{\pi V}{2 V_\pi} + \phi\right] \tag{5.28}$$

where

V_π = voltage required to achieve a π optical phase shift,
ϕ = static bias phase shift $\pm \pi/4$,
V = peak voltage of the input RF signal,
P_{pk} = peak power of the laser diode.

The peak voltage, V, of the modulating RF signal can also be expressed in terms of the modulator input impedance R_M, which is normally 50 Ω and the rms RF signal drive power S as follows

$$S = \frac{V^2}{2 R_M} \qquad (5.29)$$

Substituting (5.29) into (5.28) yields

$$P = P_{pk} \sin^2 \left[\frac{\pi \sqrt{2 R_M S}}{2 V} + \frac{\pi}{4} \right] \qquad (5.30)$$

The modulator transfer function is defined as

$$\xi_M = \left. \frac{\delta P}{\delta \sqrt{S}} \right|_{s=0} = \frac{\pi P_{pk} \sqrt{2 R_M}}{2 V_\pi} \qquad (5.31)$$

As before, the optical power transfer coefficient ξ_E from the modulator to the detector is expressed as

$$\xi_E = \xi_M \xi_F \eta_p \sqrt{R_P} \qquad (5.32)$$

and the overall electrical function from modulator to detector is ξ_E^2. Due to additional coupling loss between the optical fiber and the input-output ports of an external modulator as well as propagation loss in the optical waveguide of an external modulator the transfer coefficient ξ_F is smaller than in the case of a directly modulated link.

The electro-optic modulator is a noiseless device, but it passes the thermal noise power from the input amplifier through the transfer function. The laser also contributes noise which is expressed in terms of the RIN. The photodiode contributes shot noise. Thus, analogous to (5.16), the total noise out of an externally modulated link is

$$N_{out} = \xi_E^2 KTB + \xi_c^2 \text{RIN}_{EOM} (I_{bias} - I_{th})^2 R_L B$$
$$+ 2e(I_{bias} - I_{th}) \xi_c F_P B \sqrt{R_L R_P} \qquad (5.33)$$

where ξ_c is the optical transfer function from the laser diode terminal to the detector terminals (Figure 5.11) and is defined in a manner analogous to (5.4). That is,

$$\xi_c = \frac{P_{pk} \xi_F \eta_p \sqrt{R_P}}{2 (I_{pk} - I_{th}) \sqrt{R_L}} \qquad (5.34)$$

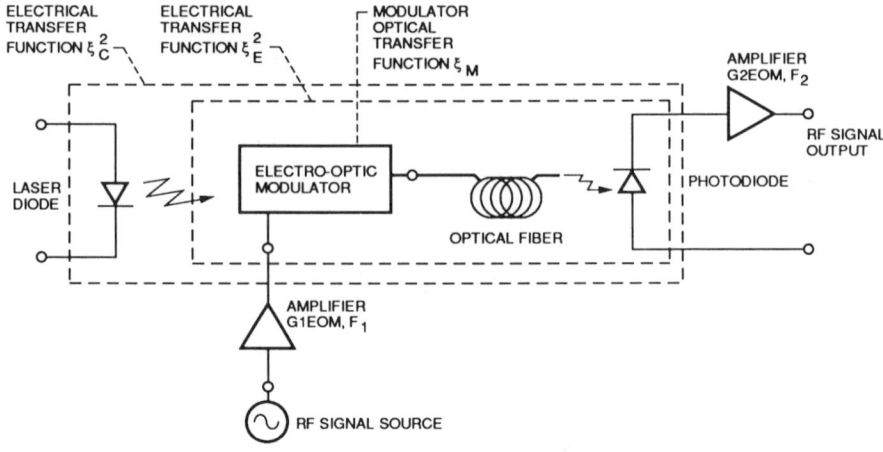

Figure 5.11 Simplified externally modulated link setup showing the various transfer functions.

The corresponding electrical transfer function is ξ_c^2; ξ_c^2 will generally have a smaller value than ξ because of additional optical losses, α, in ξ_F, and the laser power must be attenuated by some factor $y < 1$ from its peak value to prevent optical damage of the modulator. Thus,

$$\xi_c = \alpha y \xi \tag{5.35}$$

As before, the noise figure F_{EOM} of the externally modulated link can be written as

$$F_{EOM} = 1 + \frac{\text{RIN}_{EOM}(I_{\text{bias}} - I_{\text{th}})^2 R_L}{KT}\left[\frac{\xi_c^2}{\xi_M^2}\right]$$

$$+ \frac{2e(I_{\text{bias}} - I_{\text{th}}) F_P \sqrt{R_L R_P}}{\xi KT}\left[\frac{\xi_c^2}{\xi_M^2}\right]\left[\frac{\xi}{\xi_c}\right] \tag{5.36}$$

The link is normally preceded and followed by amplifiers of gain and noise figures G_{1EOM} and F_1, and G_{2EOM} and F_2, respectively. As in the previous case, the gain of the input amplifier cannot be arbitrarily selected because this would distort the signal. Hence, an optimum value for G_{1EOM} is determined as explained below. The modulation index m for an electro-optic modulator under small-signal conditions is defined as [7]:

$$m = \frac{\pi V_{\text{max}}}{V_\pi} \tag{5.37}$$

Combining (5.29), (5.31), and (5.37) yields

$$S = \frac{m^2 V_\pi^2}{\pi^2 2 R_M}$$

$$= \frac{m^2 P_{pk}^2 (\xi_F \eta_p \sqrt{R_P})}{4 \xi_E^2} \qquad (5.38)$$

Substituting (5.34) into (5.38) yields

$$S = \frac{m^2 \xi_c^2 2 (I_{pk} - I_{th})^2 R_L}{4 \xi_E^2}$$

$$= \frac{m^2 (I_{pk} - I_{th})^2 R_L}{2} \left[\frac{\xi_c}{\xi_E}\right]^2 \qquad (5.39)$$

Hence, if P_i is the input peak RF signal power to the amplifier from the RF signal source, the gain, G_{1EOM}, of the amplifier should be set equal to

$$G_{1EOM} = \frac{m^2 (I_{pk} - I_{th}) R_L}{2 P_i} \left[\frac{\xi_c}{\xi_E}\right]^2 \qquad (5.40)$$

The overall link noise figure determined from the Friis formula [10] is

$$F_{EOML} = F_1 + \frac{F_{EOM} - 1}{G_{1EOM}} + \frac{F_2 - 1}{G_{1EOM} \xi_E^2} \qquad (5.41)$$

Substituting (5.36) and (5.40) into (5.41) yields

$$F_{EOML} = F_1 + \left[\frac{2 \operatorname{RIN}_{EOM} P_i}{m^2 KT}\right] + \left[\frac{4 e P_i (R_P/R_L) F_P}{\alpha y m^2 \xi (I_{pk} - I_{th}) KT}\right]$$

$$+ \left[\frac{2 (F_2 - 1) P_i}{\alpha^2 y^2 m^2 \xi^2 (I_{pk} - I_{th})^2 R_L}\right] \qquad (5.42)$$

Figures 5.12 and 5.13 compare F_{DML} and F_{EOML} at a fixed $\operatorname{RIN}_{DM} = -130$ dB/Hz and $m = 0.3$. In Figure 5.12, RIN_{EOM} is also taken to be -130 dB/Hz and F_{EOML} is plotted for various values of αy. This corresponds to the situation where the same laser is used for both the direct and external modulation links. Clearly, direct modulation is superior. In Figure 5.13, RIN_{EOM} is taken to be -160 dB/Hz. This cor-

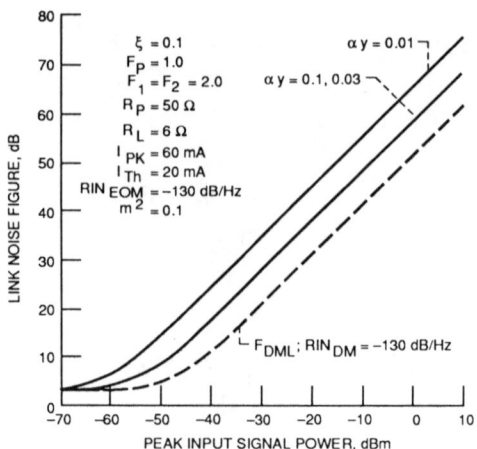

Figure 5.12 Computed overall link noise figure of an externally modulated link having a laser diode with large RIN.
Source: Yen, H.W., C.M. Gee, and H. Blauvelt, "High-Speed Optical Modulation Techniques," *Optical Technology for Microwave Applications II,* SPIE, Vol. 545, 1985, pp. 2–9. Reprinted with permission.

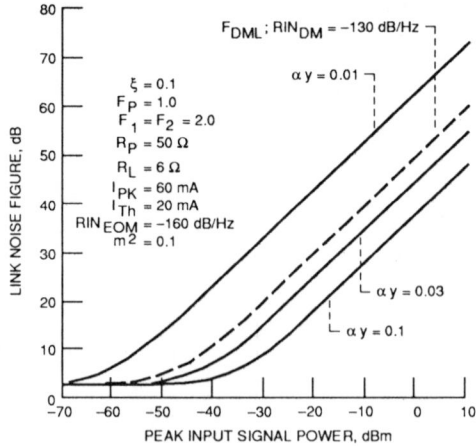

Figure 5.13 Computed overall link noise figure of an externally modulated link having a laser diode with low RIN.
Source: Yen, H.W., C.M. Gee, and H. Blauvelt, "High-Speed Optical Modulation Techniques," *Optical Technology for Microwave Applications II,* SPIE, Vol. 545, 1985, pp. 2–9. Reprinted with permission.

responds to the case where a low noise, low-bandwidth laser is used in the external modulation link. This figure shows that for small $\alpha\, y$ direct modulation is still superior, but for $\alpha\, y \approx 1$, F_{EOML} can be significantly smaller than F_{DML}.

5.5 LINK BANDWIDTH

The link bandwidth is determined by the low and the high frequency limits of the laser diode frequency response [13] for a directly modulated system. The low frequency limit is set by the cut-off frequency of the low-pass filter in the laser diode bias tee. The high frequency limit is set by the roll off in the frequency response of the laser diode, which has a typical slope of 40 dB/decade beyond the relaxation oscillation frequency. This characteristic is illustrated in Figure 2.16 of Chapter 2. Most microwave applications require a narrow bandwidth typically less than 500 MHz [13]. Commercially available laser diodes are capable of being directly modulated at frequencies of up to 10 GHz [14]. However, laboratory experimental laser diodes with a relaxation frequency as high as 30 GHz have been recently demonstrated [15].

For externally modulated links, commercially available electro-optic modulators are capable of operating up to 3 GHz [16]. However, laboratory experimental electro-optic modulators with a bandwidth as high as 40 GHz, have been recently demonstrated [17]. At the receiver end, commercially available photodiodes are capable of direct detection up to 15 GHz [14] while laboratory experimental diodes with -3 dB cut-off frequency as high as 110 GHz have been demonstrated [18]. Hence, the bandwidth requirements for most RF applications can easily be met. For single-mode optical fibers, an additional bandwidth limitation arises due to optical fiber dispersion. This will be discussed in Section 5.8.

5.6 HARMONIC AND INTERMODULATION DISTORTIONS

In the case of directly-modulated links the nonlinearity of the laser diode drive current *versus* the optical power intensity characteristic can generate higher harmonic and *intermodulation distortion* (IMD) products. Harmonics and IMD caused by this type of nonlinearity are discussed in detail in Section 8.5.1 of Chapter 8. In the case of externally modulated links the nonlinearity in the Mach-Zehnder modulator's large signal optical intensity *versus* voltage response relationship also causes harmonic and IMD, as does nonlinearity in the input amplifier.

Section 5.5 pointed out that most fiber optic links for microwave applications require a bandwidth of less than an octave, and so the higher harmonics and second-order IMD products are outside the passband and thus do not pose a problem. However the third-order IMD product frequency components are within the link passband and are therefore responsible for the spurious response. Thus, the level of third-order IMD products determines the link linearity.

The IMD is measured by combining two equal power RF signals (-10 dBm each) at closely spaced frequencies f_1 and $f_2 (f_2 - f_1 = 1$ MHz) in a power combiner, and using the composite signal to intensity-modulate the laser diode, or drive the external modulator. Because these devices are nonlinear, second-order intermodulation tones will appear at $f_1 + f_2$ and $f_2 - f_1$, third-order intermodulation tones will appear at $2f_1 - f_2$, $2f_2 - f_1$, and the harmonics at $2f_1$, $2f_2$, $3f_1$, $3f_2$, and so forth [5].

The ratio of the output amplitude to the input amplitude is called the *transfer function* of the device for the given amplitudes [19]. Plotted on a log-log scale, the fundamental transfer function is a line with a slope of unity, the second-order transfer function is a line with a slope of two, and the third-order transfer function is a line with a slope of three. Both the second- and third-order transfer functions intersect the fundamental transfer function line; these intersections are called the second- and third-order intercept points, respectively. Figure 5.14 illustrates the IMD products and the intercept points for a hypothetical microwave link. This hypothetical microwave link with a gain of 0 dB has a second-order output intercept point of $+17.5$ dBm and a third-order output intercept point of $+7.5$ dBm.

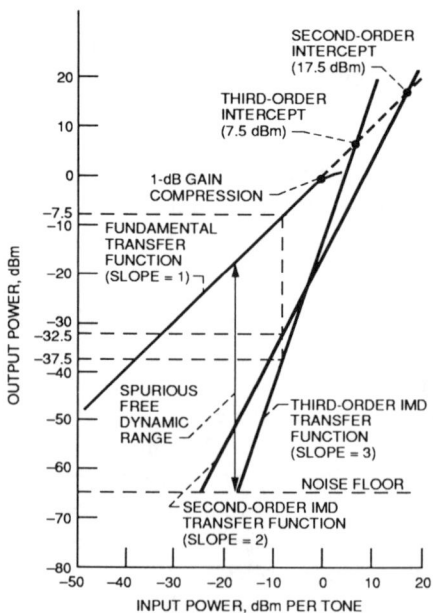

Figure 5.14 A typical illustration of second- and third-order intermodulation product levels, intercept points, noise floor, and spurious free dynamic range for a hypothetical microwave fiber optic link.

The second-order output intercept point (IP$_2$) and the third-order output intercept point (IP$_3$) can be calculated if the power in each of the components is known using the relation [19]:

$$\text{IP}_2(\text{dBm}) = 2P_f - P_2 \tag{5.43}$$

$$\text{IP}_3(\text{dBm}) = (1/2)(3P_f - P_3) \tag{5.44}$$

where

P_f = output power of each of the two fundamental tones in dBm,
P_2 = output power in each of the second-order IMD products in dBm,
P_3 = output power in each of the third-order IMD products in dBm.

As a numerical example, if $P_f = -7.5$ dBm, $P_2 = -32.5$ dBm and $P_3 = -37.5$ dBm, then IP$_2$ = 17.5 dBm and IP$_3$ = 7.5 dBm. These power levels are also illustrated in Figure 5.14.

Finally, we mention that if two microwave fiber optic links are cascaded, the intercept point of the composite system is different from the intercept point of the individual links. The composite output third-order intercept is given by [19]:

$$T_3\,(\text{dBm}) = D - 10 \log\left[1 + \frac{E}{(G\,F)}\right] \tag{5.45}$$

where

D = third-order output intercept of the second link in dBm;
E = third-order output intercept of the second link in mW;
F = third-order output intercept of the first link in mW;
G = gain of the second link, which is dimensionless.

As a numerical example, if the third-order intercept of the first and the second link are 30 and 40 dBm, respectively, and the gain of the second link is 0 dBm, the composite output third-order intercept T_3 is determined as follows:

$$D = 40 \text{ dBm}$$
$$E = \text{antilog}(40/10) = 10000 \text{ mW}$$
$$F = \text{antilog}(30/10) = 1000 \text{ mW}$$
$$G = 1$$
$$T_3 = 29.6 \text{ dBm}$$

The 1 dB compression point shown in Figure 5.14 is a measure of the output power level, when the input power has been raised to a point on the transfer characteristic that results in a 1 dB reduction in output power. This is a good indication that the nonlinear effects of saturation will begin causing distortion very rapidly with increased input signal power.

5.7 SPURIOUS-FREE DYNAMIC RANGE AND MAXIMUM SIGNAL-TO-NOISE RATIO

The discussion presented in this section is based on [5]. The dynamic range of the fiber optic link is a measure of the variation of signal levels that can be carried by the link. The dynamic range is typically defined as the ratio of the fundamental output to the third-order IMD products. Since the IMD level decreases faster than the fundamental, reduction of the input signal level yields any value of dynamic range required. However, as the input signal level decreases, the signal-to-noise ratio also decreases because the noise out of the link is constant for a specific noise bandwidth. Thus, there is an input signal level at which the IMD level matches the noise floor, and at that input level the dynamic range is maximized. Hence, the *spurious-free dynamic range* (SFDR) is the level of IMD suppression achieved when that level equals the link signal-to-noise ratio. The link-noise floor can be directly determined either from (5.16) or (5.33). Alternatively, if the link insertion loss and noise figure are known, (5.17) can be used to determine the link noise floor. In Figure 5.14, the noise level of a hypothetical link in a 1 MHz noise bandwidth and the SFDR are also indicated. The SFDR can be calculated, if the noise figure of the link is known, using the relation [19]:

$$\text{SFDR (dB)} = (2/3) [IP_3 - NF + 174 - 10 \log(B)] \qquad (5.46)$$

where

IP_3 = third-order output intercept point in dBm,
B = link bandwidth in Hz,
NF = link-noise figure in dB.

As a numerical example, if $IP_3 = 7.5$ dBm, $B = 1$ MHz and NF = 49.5 dB, the SFDR = 48 dB.

5.8 SINGLE-MODE OPTICAL FIBER DISPERSION

The dispersion in a single-mode optical fiber is caused by both material and waveguide dispersion [20]. The variation of the refractive index of the core material with wavelength causes material dispersion. Material dispersion causes the speed of light in the fiber to be different for each wavelength. Thus, a narrow pulse after propagating through the fiber broadens at the output. This pulse broadening, τ_{mat} can be approximated by [20]:

$$\tau_{mat} = \frac{dt_{mat}}{d\lambda} \sigma_\lambda = -\frac{L}{c} \lambda \frac{d^2n}{d\lambda^2} \sigma_\lambda \qquad (5.47)$$

where

t_{mat} = group delay,
λ = wavelength of the optical source,
σ_λ = rms spectral width of the optical source,
L = total length of the fiber,
c = 3×10^8 m/s,
n = refractive index of the fiber core.

In the above equation, the factor

$$\frac{\lambda}{c} \cdot \frac{d^2n}{d\lambda^2}$$

is the material dispersion.

The variation in the modal propagation constant β with the ratio a/λ, where a is the core radius and λ is the wavelength, causes waveguide dispersion. Waveguide dispersion causes the speed of light in the fiber to attain an effective value which is in between the velocities in the core, and cladding materials. Insofar as the refractive index is a function of the wavelength, this type of dispersion also causes pulse broadening. This pulse broadening, τ_{wg} can be approximated by [20]:

$$\tau_{wg} = -\frac{0.003 \, \sigma_\lambda L}{c \lambda} \qquad (5.48)$$

Both material and waveguide dispersion are measured in picoseconds (of pulse spreading) per nanometer (of source spectral width) per kilometer (of path length), (ps/(nm·km)). The sum of these two dispersions is the total dispersion for a single-mode fiber.

Figure 5.15 shows the material dispersion, and the waveguide dispersion of a typical single-mode, fused-silica-core fiber. The material dispersion is observed to dominate over the waveguide dispersion. Further, the material dispersion and the waveguide dispersion have opposite signs and therefore completely cancel at a certain wavelength. The zero dispersion wavelength occurs at about 1.27 μm for pure silica. The attenuation characteristic of the single-mode optical fiber takes a dip at

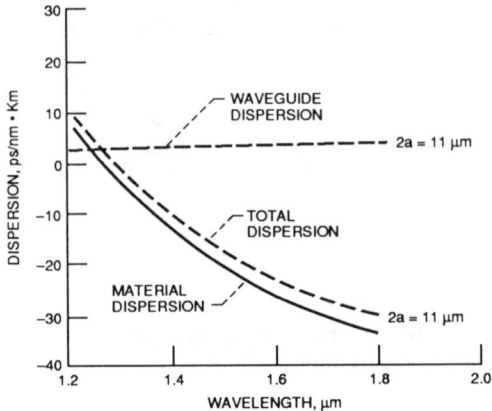

Figure 5.15 Magnitude of material and waveguide dispersion as a function of optical wavelength for a single-mode fused-silica-core fiber.
Source: Keiser, G., *Optical Fiber Communications,* New York, McGraw-Hill, 1983, p. 67. Reprinted with permission.

about 1.3 μm and 1.55 μm. Therefore, to shift the point of zero dispersion from 1.27 μm to either 1.3 or 1.55 μm would be advantageous. This is achieved by employing a special radial refractive index profile, such as the one in which the index of the inner cladding next to the core is lower than that of the outermost cladding. Such a profile is known as the *depressed cladding profile,* and the type of fiber is known as the *dispersion shifted fiber.* By employing a more complicated radial refractive index profile, we can achieve very low dispersion over a range of wavelengths between 1.3 and 1.55 μm. This type of optical fiber is known as *dispersion flattened fiber.*

If D_t is the total dispersion of an optical fiber, the pulse spread $\Delta\tau$ is [20]:

$$\Delta\tau = L\, D_t\, \sigma_\lambda \tag{5.49}$$

The bandwidth of the fiber is approximately given by [20]:

$$\Delta f = \frac{0.44}{\Delta\tau}$$

Substituting from (5.49) yields

$$\Delta f = \frac{0.44}{L\, D_t\, \sigma_\lambda} \tag{5.50}$$

This results in a bandwidth-distance product of

$$\Delta fL = \frac{0.44}{D_t \sigma_\lambda} \quad (5.51)$$

Commercially available single-mode fiber have a typical value of dispersion of about 3.5 ps/(nm·km) at 1.3 μm wavelength [21]. This results in a bandwidth-distance product of $(0.44/3.5 \times 10^{-12}) = 125$ GHz·km.

5.9 OPTICAL LOSSES IN A MICROWAVE FIBER OPTIC LINK

5.9.1 Single-Mode Optical Fiber Attenuation

Absorption and scattering of light propagating through a silicon optical fiber leads to attenuation [22]. Attenuation is expressed in decibels per kilometer (dB/km) and is a function of the operating wavelength. At lower wavelengths, typically below 1.5 μm, the absorption losses are due to an ultraviolet absorption band edge, which results from resonances associated with the electronic structures of the crystal atoms. Above the band edge, Rayleigh scattering dominates the loss mechanism. This takes place because of the microscopic variations around the average material density, and also local microscopic variations in the composition. Each of these causes fluctuations of refractive index and consequently scatters the incident optical power. The scattered power and, hence, the attenuation are inversely proportional to the fourth power of the wavelength. Experiments show that Rayleigh scattering, rather than the ultraviolet absorption band edge, is the main cause of loss in silicon fibers at wavelengths of less than 1.5 μm. At wavelengths higher than 1.5 μm, the absorption loss is due to infrared absorption caused by resonances associated with the lattice vibrations of the atoms. Thus, in the valley formed by the attenuation curves due to Rayleigh scattering and infrared absorption lies the region of minimum attenuation.

Figure 5.16 shows the attenuation characteristic of an ultra-low-loss germanosilicate (GeO_2) single-mode optical fiber. The absorption peak at 1.39 μm is due to hydroxyl impurity in the glass which is typically few parts per billion (ppb). However, it is possible to reduce the hydroxyl content to less than 0.8 ppb by resorting to *vapor-phase axial deposition* (VAD) process for the manufacture of optical fibers. In the case of optical fibers fabricated by the VAD process, the absorption peak at 1.39 μm is barely present, and the attenuation is less than 1.0 dB/km over the wavelength range of 1.1 to 1.7 μm [23]. Finally, the typical attenuation of a single-mode dispersion shifted optical fiber is about 0.35 and 0.20 dB/km at 1.3 and 1.55 μm wavelength, respectively.

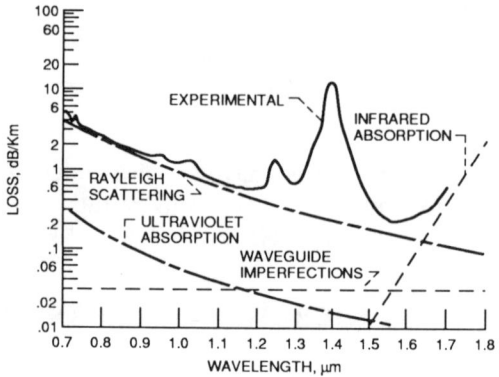

Figure 5.16 Measured attenuation of a single-mode optical fiber.
  Source: Li, T., "Structures, Parameters, and Transmission Properties of Optical Fibers," *Proc. IEEE,* Vol. 68, No. 10, October 1980, pp. 1175–1180. Reprinted with permission.

5.9.2 Laser Diode to a Single-Mode Fiber Coupling Loss

The *numerical aperture* (NA) of a single-mode fiber is given by the relation:

$$\text{NA} = n_1 \sin\theta = (n_1^2 - n_2^2)^{1/2} \tag{5.52}$$

where θ is the internal half-cone angle of acceptance, n_1 and n_2 are the refractive index of the cone and the cladding, respectively. A typical single-mode fiber with a core diameter of 8 to 10 μm has a numerical aperture of 0.08. The internal half-cone angle of acceptance, θ, is then 4.6°.

The emission pattern of semiconductor laser diodes has a divergence angle of 30 to 50° in the plane perpendicular to the junction, and a divergence angle of 5 to 10° in the plane parallel to the junction. Thus, an asymmetrical optical system, such as a cylindrical lens is required for coupling a laser diode to a single-mode optical fiber [24]. The coupling optics must form a beam which focuses to a spot of diameter less than the core diameter of the fiber, and also have a convergence angle, θ, less than that derived from the numerical aperture of the fiber. This type of coupling results in a loss of about 2 dB.

5.9.3 Connector Loss

Connector loss is a function of the physical alignment of one fiber core to another. The optical power loss of a connector is typically between 0.7 to 1.0 dB, depending on the style of the connector and quality of manufacture [25].

5.9.4 Splice Loss

Two fibers may be joined in a permanent fashion by fusion, welding, chemical bonding, or mechanical joining. As a consequence, the splice loss that is introduced into the system is about 0.2 dB [26].

5.9.5 Coupler Loss

Interconnecting components such as directional couplers and star couplers also introduce excess loss into the fiber optic link. Directional couplers are used to couple a desired fraction of the optical signal from one single-mode fiber to another, preserving propagation direction. The coupler is fabricated by removing the cladding material from the fiber and placing the core sufficiently close, thus allowing the signal to couple from one fiber to another by the evanescent fields. Typical excess loss of the coupler is in the range of 0.07 dB to 0.75 dB depending upon the coupling ratio tolerance and thermal stability that is required [27].

Star couplers allow the transmitter to communicate with several receivers simultaneously. The coupler is fabricated by forming 3 dB fused coupling regions between appropriate fiber pairs. Typical excess loss of a 1 × 8 star coupler is in the range of 1.3 to 4.0 dB, depending on the coupling tolerance and thermal stability that is required [28].

5.10 PERFORMANCE COMPARISON OF MICROWAVE FIBER OPTIC LINKS

The performance characteristics of a typical directly modulated microwave fiber optic link, and an externally modulated link are compared in Table 5.2. Direct modulation offers simplicity, low drive power, and less overall link loss. On the other hand, the use of external modulators relaxes the need for laser diodes with a large separation in frequency between the signal band edge, and the onset of relaxation oscillations. The disadvantages of the externally-modulated link are: additional coupling loss, higher drive power requirements, and more distortion. Both of these techniques need to be evaluated for their applicability to a specified system requirement.

5.11 FIBER OPTIC LINK DESIGN EXAMPLES

5.11.1 Directly Modulated, Medium Bandwidth, Long Distance, Point-To-Point Microwave Link

Let us suppose that we require to have an RF fiber optic link with a laser diode

Table 5.2
Performance Comparison of Directly Modulated and Externally Modulated
Microwave Fiber Optic Links

	Directly Modulated	Externally Modulated
Laser Transmitter Type	1.3 μm, InGaAsP, BH	1.3 μm, InGaAsP, BH
Photodiode Receiver Type	PIN, InGaAs	PIN, InGaAs
Modulator Type	—	Mach-Zehnder on $LiNbO_3$
Fiber Type and Length	single-mode, 1.1 km	single-mode, 1 km
Laser RIN (dBm/MHz)	−59.7	−66.0
Bandwidth (GHz)	4.1 to 4.7	2.0 to 12.0
Input Level (dBm)	0	0
Signal-to-Noise Ratio at Output in 1 MHz Bandwidth (dB)	59.3	53
Link Loss (dB)	36.4	67
Dynamic Range at Output in 1 MHz Bandwidth (dB)	44	—
Third-Order Intermodulation Product Suppression (dBc)	−36	—
Insertion Loss of Modulator (dB)	—	7
Voltage for 100% Modulation (V rms)	—	7.6
Reference	[5]	[5]

transmitter at point A, and a photodiode receiver at point B, as shown in Figure 5.17. Further, let us suppose that the distance of separation between points A and B is 2.4 km, and the bandwidth of the signal to be transmitted is 500 MHz. This type of link would be required, for instance, in X-band doppler radars which usually require less than 500 MHz bandwidth for transmitting the down-converted signal to the processor.

In designing the link, the assumption made is that the link noise is dominated by laser noise, and not by the receiver noise. This is a valid assumption as the transmission distance is limited mainly by fiber bandwidth instead of attenuation,

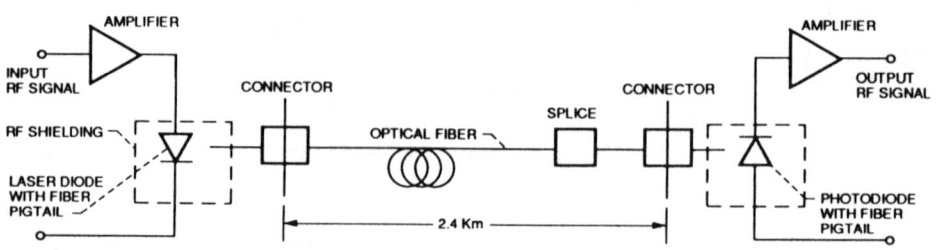

Figure 5.17 A typical design example: directly modulated optic link.

and as a result the optical power at the receiver is usually quite high. The relaxation oscillation frequency of the laser at the operating bias current should be at least four times higher than the signal band edge. This would ensure that the RIN is sufficiently small.

Let us suppose that the laser chosen for the link is an InGaAsP laser, operating at a wavelength of 1.3 μm, with RIN of −140 dB/Hz at 0.5 GHz and 40.0 mA bias current. Further, let the threshold current of the laser be 20 mA. The laser diode is assumed to be impedance matched to a 50 Ω line and also has a fiber pigtail. A single-mode fiber with core and cladding diameters of 9 and 125 μm, respectively, is selected as the transmission medium. This optical transmission medium has low attenuation and almost zero dispersion at 1.3 μm wavelength. Hence, optical power loss is a minimum and the bandwidth virtually infinite.

Most commercial fibers come in lengths of 2 km or more [21]. Let us therefore suppose that one splice is needed to obtain a length of 2.4 km. Further, for interconnection, a connector is needed at either ends of the fiber. Based on the above information and referring to Section 5.9, the link optical loss budget is computed as follows:

- One splice at 0.2 dB each   0.2 dB
- Two connectors at 1.0 dB each   2.0 dB
- 2.4 km of single-mode optical fiber at 0.5 dB/km attenuation   1.2 dB
- Coupling loss between fiber and photodiode   2.0 dB
- Loss due to aging, misalignment   2.0 dB
- Total link optical loss   7.4 dB

Let us suppose that the laser diode has a fiber pigtail and the differential quantum efficiency (η_L) as determined at the end of the fiber, is 30% and the diode resistance R_L, is 6 Ω. Further, let us suppose that an InGaAs photodiode with a responsivity of 0.7 mA/mW and resistance R_P of 50 Ω is used as a detector. Then, the laser diode and the photodiode power conversion efficiencies and the optical power transfer efficiency as determined from (5.4) are

$$\eta_L/\sqrt{R_L} = 0.3/\sqrt{6} = 0.122$$

$$\xi_F = \text{antilog}(-7.4/10) = 0.182$$

$$\eta_P \sqrt{R_P} = 0.7 \sqrt{50} = 4.950$$

The overall link transfer function ξ as determined from (5.4) is

$$\xi = 0.11$$

The link noise figure as computed by using (5.18) is

$$F_{DM} = 37.8 \text{ dB}$$

Amplifiers are normally used to amplify the RF signal before intensity-modulating the laser diode, and also before processing the detected signal at the output of the photodiode. Their noise figures can also be taken into consideration and the overall link noise figure can be determined. Let us suppose that the peak input RF signal level before amplification is -10 dBm. Then, the gain of the amplifier at the input to the link as determined from (5.26) is

$$G_{1DM} = 10.8 \text{ dB}$$

Further, let us suppose that the noise figures of the amplifiers mentioned above are 2.0 dB. Then the overall link noise figure as determined by using (5.27) is

$$F_{DML} = 27 \text{ dB}$$

5.11.2 Directly Modulated, Narrowband, Short-Distance, Multiport Microwave Link

This type of link would be required, for instance, in an active phased array antenna system in which several slave oscillators were subharmonically injection-locked to a master oscillator. Figure 5.18 schematically illustrates the link configuration. A star coupler serves as a $1:N$ power divider. The length of the fiber required to link the master oscillator with a slave oscillator *via* the star coupler is indicated in Figure 5.18. Suppose that the frequency of the synchronizing RF signal is 1 GHz, and the locking bandwidth of the oscillator is 50 MHz. Further, let N be equal to 16. The

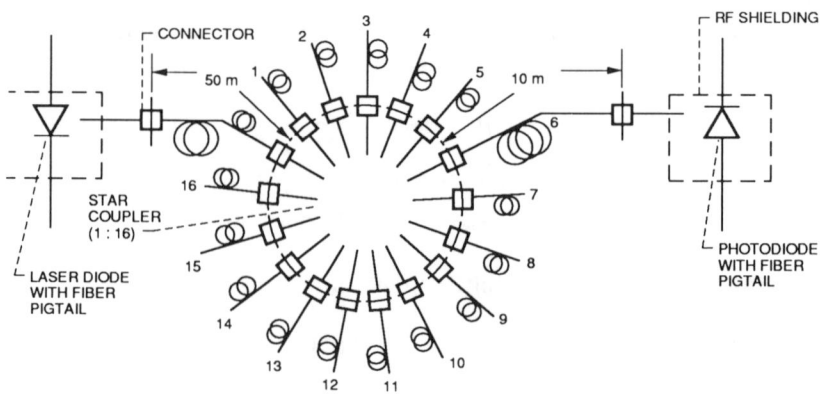

Figure 5.18 A typical design example: directly modulated fiber optic link with N outputs.

link optical loss budget is computed as follows:

- 60 m of fiber at 0.5 dB/km   0.03 dB
- Four connectors at 1.0 dB each   4.00 dB
- Division loss = $1/N = 1/16$ 12.00 dB
- Excess loss of star coupler   2.00 dB
- Coupling loss at photodiode   2.00 dB
- Aging, misalignment   3.00 dB
- Total loss from source to one of the slave stations   23.03 dB

The optical fiber power transfer function ξ_F is antilog($-23.03/10$), and is equal to 0.005. The rest of the link analysis proceeds as in the previous example.

5.11.3 Point-to-Point Microwave Link When Limited by Receiver Noise

Let us suppose that we require a 20 km RF fiber optical link. In this case, the link noise is determined by the receiver noise instead of the laser noise, since the optical power at the receiver is usually very small. For this purpose let us suppose that we have selected a laser diode with a fiber pigtail, and the power output at the end of the fiber pigtail is -10 dBm. Further, suppose that the total itemized optical power loss as in the previous examples turns out to be 21.2 dB. Then the power incident on the photodiode is -31.2 dBm. If the photodiode has a responsivity of 0.7 mA/mW, then the photocurrent I_P at the output of the photodiode is 0.531 μA. Let the bandwidth of the signal transmitted be 250 MHz. The signal power at the output is proportional to I_P^2, whereas the short noise power is proportional to $2eI_P\Delta f$. The electronic charge e is equal to 1.6×10^{-19}C and Δf is the bandwidth, which is equal to 250×10^6 Hz. The signal-to-noise ratio at the photodiode therefore is

$$(S/N)_{in} = I_P/2e\Delta f$$
$$= 38.2 \text{ dB}$$

If a postdetection amplifier of noise figure 3 dB with a gain of 12 dB is used, the overall signal-to-noise ratio is

$$(S/N)_{out} = 35.2 \text{ dB}$$

5.11.4 Externally Modulated, Short-Distance, Point-to-Point Microwave Link

The advantage of an externally modulated link is that it is not limited in frequency by the laser relaxation resonance and the associated RIN. Hence, a laser diode with a low RIN can be selected. Let us suppose that the distance of separation between the laser diode and the photodiode be 1 km, and the bandwidth of the signal to be transmitted is 500 MHz at a center frequency of 12 GHz. Further, the laser diode, the photodiode and the electro-optic modulator are assumed to have the following typical characteristics:

$$RIN_{EOM} = -160 \text{ dB/Hz}$$
$$R_L = 6 \ \Omega$$
$$\eta_L = 0.3 \text{ mW/mA}$$
$$\lambda = 1.3 \ \mu\text{m}$$
$$I_{pk} = 32 \text{ mA}$$
$$I_{bias} = 24 \text{ mA}$$
$$I_{th} = 16 \text{ mA}$$
$$P_{pk} = 100 \ \mu\text{W}$$
$$m = 0.1$$
$$\eta_P = 0.7 \text{ mA/mW}$$
$$R_P = 50 \ \Omega$$
$$R_M = 50 \ \Omega$$
$$V_\pi = 7.0 \text{ V}$$

The link loss budget is calculated as follows:

- 1 km of single-mode optical fiber at 0.5 dB/km attenuation 0.5 dB
- Coupling loss between the laser diode and the external modulator 2.0 dB
- Insertion loss of the modulator 5.0 dB
- Coupling loss between the modulator output port and the optical fiber 2.0 dB
- Coupling loss between the fiber and the photodiode 2.0 dB
- Total link optical loss 11.5 dB

The optical path transfer function ξ_F is antilog($-11.5/10$) and is equal to 0.071. The modulator transfer function ξ_M, as determined from (5.31), is

$$\xi_M = 2.244 \times 10^{-4}$$

The laser diode and the photodiode power conversion efficiencies as determined from (5.4) are

$$\eta_L/\sqrt{R_L} = 0.3/\sqrt{6} = 0.1225$$
$$\eta_P\sqrt{R_P} = 0.7\sqrt{50} = 4.9497$$

The transfer functions ξ, ξ_E, and ξ_C as determined from (5.4), (5.32), and (5.34) are

$$\xi = 0.0431$$
$$\xi_E = 7.8861 \times 10^{-5}$$
$$\xi_C = 4.4834 \times 10^{-4}$$

The noise figure F_{EOM} of the externally modulated link as determined from (5.36) is

$$F_{EOM} = 49.8 \text{ dB}$$

5.11.5 Computer Aided Design Modeling of a Laser Diode and a Photodiode for a Microwave Fiber Optic Link

This example is based on the work reported in [29]. Let us suppose that we wish to obtain equivalent circuit models of a laser diode and a photodiode, which are later to be used in a microwave fiber optic link. The laser diode being a forward bias device is modeled as a parallel RC circuit. Each element in the equivalent circuit model has to have its origin within the device. For example, the junction resistance of a forward biased laser diode can be represented by a resistance R_L. The diffusion capacitance can be represented as a capacitance C_L. In addition, the parasitic elements due to contacts and bond wire can be represented by resistances R_1 and R_2 and inductance L_B. The resulting small-signal, lumped-element, bias-dependent equivalent circuit model of the laser diode is shown in Figure 5.19(a).

The elements of the equivalent circuit model are optimized over the frequency range of operation of the link to match, the measured reflection coefficient (magnitude and angle of S_{11}) using EEsof's Touchstone circuit analysis and optimization programs [30]. As an example, the optimized element values over the frequency range of 2 to 12 GHz for a commercially available GaAlAs laser diode [29] are also presented in Figure 5.19(a).

The photodiode being a reverse bias device is modeled (Figure 5.19(b)) as a series RC circuit, with R_P being the contact resistance, and C_P being the depletion capacitance. A current source of magnitude ηI_L in parallel with the depletion capacitance accounts for the behavior of the photodiode as a generator of photocurrent when illuminated. The photocurrent is assumed to be directly proportional to the optical illumination. As explained in Section 5.2, the transfer function η accounts for all the optical losses in the link, as well as the conversion efficiencies of the

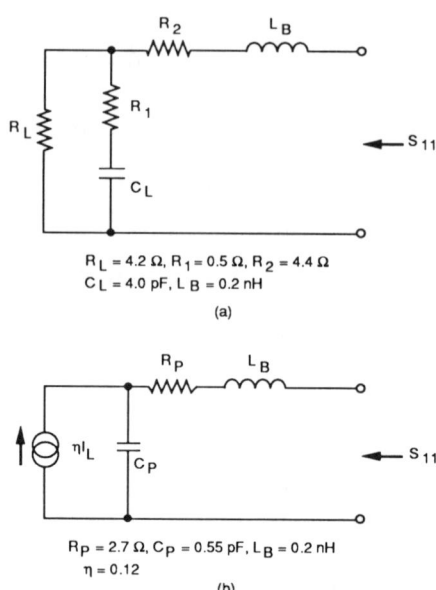

Figure 5.19 A typical CAD example: small signal lumped element equivalent circuit model of a semiconductor laser diode and a Schottky photodiode.
Source: After Hsu, H.-P., M. de La Chapelle, and J.J. Gulick, "Fiber Optic Links for Microwave Signal Transmission," *High Frequency Optical Communications,* SPIE, Vol. 716, 1986, pp. 69–75.

laser diode and the photodiode. The resulting small-signal, lumped-element equivalent circuit model of a Schottky photodiode, together with the optimized element values [29] over a frequency range of 2 to 12 GHz, is also shown in Figure 5.19(b). Finally, we mention that the performance of the microwave link predicted by using the above models are observed to be in good agreement with measurements [29].

REFERENCES

1. Belden Wire and Cable, Master Catalog 885, Section VIII, Fiber Optic Cable, p. 159.
2. Waveguide, *Bends and Twists Data Sheet,* Micro Bends Corp., CT.
3. Fink, D.G., and D. Christiansen, *Electronics Engineers' Handbook,* 2nd Ed., New York, McGraw-Hill, 1982, Section 9, p. 9.
4. Precision Tube Co., Inc., Coaxitube Semi-Rigid Coaxial Cable Catalog, p. 22.
5. Stephens, W.E., and T.R. Joseph, "System Characteristics of Direct-Modulated and Externally Modulated RF Fiber-Optic Links," *IEEE J. Lightwave Technol.,* Vol. LT-5, No. 3, March 1987, pp. 380–387.
6. Ito, T., S. Machida, K. Nawata, and T. Ikegami, "Intensity Fluctuations in Each Longitudinal Mode of a Multimode AlGaAs Laser," *IEEE J. Quantum Electronics,* Vol. QE-13, No. 8, August 1977, pp. 574–579.

7. Yen, H.W., C.M. Gee, and H. Blauvelt, "High-Speed Optical Modulation Techniques," *Optical Technology for Microwave Applications II*, SPIE, Vol. 545, 1985, pp. 2–9.
8. Koyama, F., T. Tanbun-ek, S. Arai, S. Wang, Y. Suematsu, and K. Furuya, "Suppression of Intensity Fluctuation of a Longitudinal Mode in Directly Modulated GaInAsP/InP Dynamic Single-Mode Laser," *Electronics Letters*, Vol. 19, No. 9, April 1983, pp. 325–327.
9. Way, W.I., R.S. Wolff, and M. Krain, "A 1.3-μm 35-Km Fiber-Optic Microwave Multicarrier Transmission System for Satellite Earth Stations," *IEEE J. Lightwave Technol.*, Vol. LT-5, No. 9, September 1987, pp. 1325–1331.
10. Friis, H.T., "Noise Figures of Radio Receivers," *Proc. IRE*, Vol. 32, No. 7, July 1944, pp. 419–422.
11. Toshiba Corp., Fiber-Optic Semiconductor Devices Data Sheet, Model TOLD 332S/TOLD 333S (1.3 μm), Tokyo.
12. Ortel Corp., High Speed Fiber-Pigtailed GaAlAs Laser Diodes and Transmitters Data Sheet (SL-Series).
13. de la Chapelle, M., and H.P. Hsu, "Characterization of Fiber-Optic Links for Microwave Signal Transmission," *Optical Technol. for Microwave Applications III*, SPIE, Vol. 789, 1987, pp. 32–39.
14. Browne, J., "Window Boosts Laser Diode Power, Ortel Corp. Product Technology," *Microwaves & RF*, Vol. 24, No. 6, June 1985, pp. 159–160.
15. Uomi, K., T. Mishima, and N. Chinone, "Ultrahigh Relaxation Oscillation Frequency (up to 30 GHz) of Highly p-doped GaAs/GaAlAs Multiple Quantum Well Lasers," *Applied Physics Letters*, Vol. 51, No. 2, July 1987, pp. 78–80.
16. Optical Guided-Wave Mach-Zehnder Modulators Data Sheet, Crystal Technology, Inc.
17. Dolfi, D.W., M. Nazarathy, and R.L. Jungerman, "40 GHz Electro-Optic Modulator with 7.5V Drive Voltage," *Electronic Letters*, Vol. 24, No. 9, April 1988, pp. 528–529.
18. Parker, D.G., P.G. Say, A.M. Hansom, and W. Sibbett, "110 GHz High-Efficiency Photodiodes Fabricated from Indium Tin Oxide/GaAs," *Electronic Letters*, Vol. 23, No. 10, May 1987, pp. 527–528.
19. Jacobi, J.H., "IMD: Still Unclear After 20 Years," *Microwaves & RF*, Vol. 25, No. 11, November 1986, pp. 119–126.
20. Keiser, G., *Optical Fiber Communications*, New York, McGraw-Hill, 1983, pp. 62–67 and p. 218.
21. "At-A-Glance Guide to Optical Fiber and Cable," *Photonics Spectra*, Vol. 19, No. 9, September 1985, pp. 91–100.
22. Li, T., "Structures, Parameters, and Transmission Properties of Optical Fibers," *Proc. IEEE*, Vol. 68, No. 10, October 1980, pp. 1175–1180.
23. Hanawa, F., S. Sudo, M. Kawachi, and M. Nakahara, "Fabrication of Completely OH-Free V.A.D. Fibre," *Electronics Letters*, Vol. 16, No. 18, August 1980, pp. 699–700.
24. Saruwatari, M., and K. Nawata, "Semiconductor Laser to Single-Mode Fiber Coupler," *Applied Optics*, Vol. 18, No. 11, June 1979, pp. 1847–1856.
25. GTE Type PC-FC Single-Mode Fiber Optic Connector Data Sheet, GTE, PA.
26. UV Curved Optical Fiber Splice Data Sheet, Norland Products, NJ.
27. Single-Mode Fused Waveguide Independent Couplers (WIC) Data Sheet, Gould, Fiber Optics Operation, MD.
28. Single-Mode Tree and Star Couplers Data Sheet, Gould, Inc., Fiber Optic Operation, M.D.
29. Hsu, H.P., M. de la Chapelle, and J.J. Gulick, "Fiber Optic Links for Microwave Signal Transmission," *High Frequency Optical Communications*, SPIE, Vol. 716, 1986, pp. 69–75.
30. EEsof Touchstone Reference Manual, Version 1.5, EEsof, Inc., March 1987.

Chapter 6
Optoelectronic Switching and Gating

6.1 INTRODUCTION

An optoelectronic microwave switch or gate is constructed from planar transmission lines, such as microstrip. The dielectric material that is used in these transmission lines is a semiconductor, such as silicon or gallium arsenide. An opening in the transmission line exposes the semiconductor to optical illumination. When the optical illumination is such that its photon energy is greater than the semiconductor bandgap, a significant amount of the incident energy is absorbed. This absorbed energy creates free electron hole pairs, and creates a semiconductor plasma of a concentration that decays exponentially with depth and wavelength. Thus, the conductivity as well as the dielectric properties of the semiconductor below the optical footprint are altered. These changes in the basic properties of the semiconductor are exploited in the switching and gating of microwave signals.

Optoelectronic microwave switches have several advantages over conventional GaAs MESFET or PIN diode switches. The optoelectronic switches promise faster rise and fall time (typically in the picosecond range), broader bandwidth, ability to handle high power, and simplicity of operation. In addition, the controlling optical signal and the gated microwave signal are totally electrically isolated, and hence the switch activation is jitter-free. The optical and electrical characteristics of a microstrip optoelectronic switch is discussed in the next section. Further, the characteristics of this switch are compared with those of conventional electronic switches fabricated by using MESFETs and PIN diodes.

6.2 MICROSTRIP OPTO-ELECTRONIC SWITCH WITH TOP-SIDE EXCITATION

6.2.1 Principle of Operation

A specific example of an optoelectronic microstrip switch that utilizes the photoconductivity produced by optical illumination is illustrated in Figure 6.1. The switch consists of a 50 Ω microstrip transmission line structure fabricated on a high resistivity silicon substrate. In the top strip conductor of the microstrip, a gap of length L is etched by using conventional photolithographic techniques. When the gap is illuminated, absorption of the photons takes place in the silicon substrate. An absorbed photon causes an electron to transfer from the bound state to the free state, thereby creating a free electron-hole pair. This is an intrinsically fast process and requires only the wavelength of the illumination to be within the absorption band of the semiconductor that constitutes the substrate. Because these electronic transitions are only a few eV wide, the transfer time is as short as 10^{-15} s, or one optical cycle. Hence, the width of the transition gap does not limit the rise time of the photoconductivity. However, the amplitude and duration of the optical illumination does influence the rise and fall time, insofar as the photoconductivity increases linearly with the optical pulse energy.

The optical absorption constant, α [1], for silicon is shown as a function of the wavelength of the incident illumination in Figure 6.2. This figure suggests that if the gap in the microstrip is illuminated by an optical pulse whose wavelength is less than 0.55 μm, the switch can be turned on. In this wavelength range the absorption constant is very large, and hence the absorption depth $(1/\alpha)$ is very small. Therefore, the illumination creates a surface plasma. This surface plasma enhances the surface photoconductivity and bridges the gap in the strip conductor, thus resulting in negligible attenuation of the microwave signal as it propagates from the input to the output.

The gap length L is chosen such that it is small compared to the guide wavelength of the microwave signal ($L < \lambda_g/50$). Given this approximation, and for the purpose of analysis, the gap is represented by a lumped equivalent circuit model. The equivalent circuit model chosen for the switch in the on-state is a series conductance G_1, as shown in Figure 6.3(a). G_1 represents the surface photoconductivity of the plasma.

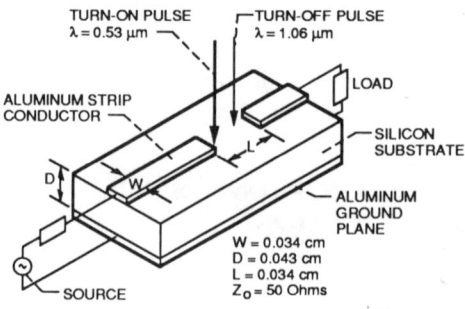

Figure 6.1 Schematic illustrating a microwave optoelectronic switch.

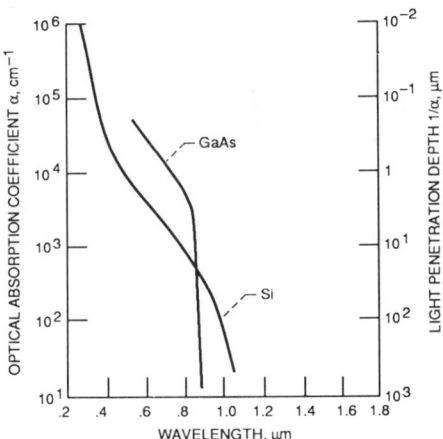

Figure 6.2 Optical absorption coefficient for silicon and GaAs as a function of the wavelength. *Source:* Sze, S.M., *Physics of Semiconductor Devices*, 2nd Ed., New York, John Wiley and Sons, 1981, p. 750. Reprinted with permission.

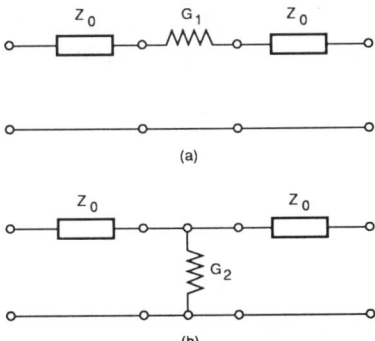

Figure 6.3 Equivalent circuit of the switch: (a) on-state; (b) off-state.

If a second optical pulse at a wavelength greater than 1 μm is made to illuminate the gap in the microstrip, the switch is turned off. In this case, the optical absorption constant as obtained from Figure 6.2 is small and the corresponding absorption depth is large. Hence, the optical illumination gives rise to volume photoconductivity. This bulk photoconductivity short circuits the top strip conductor to the bottom ground plane of the microstrip. The short circuit reflects the incident signal back to the generator resulting in negligible transmission in the forward direction. The equivalent circuit model of the switch in the off-state is represented by a shunt conductance G_2, and is shown in Figure 6.3(b). G_2 represents the volume photoconductivity of the plasma.

6.2.2 Gap Series Conductance

An expression for the gap series conductance G_1 in the on-state is presented in this section. The assumption made here is that the recombination and trapping of electronics and holes are small and hence, can be neglected. This is a valid assumption because the recombination time of electrons and holes in silicon is on the order of 1 μs, which is large compared to the picosecond optical gating period. By integrating the conductivity over the exponential profile induced by the absorbed optical intensity, an expression for the total conductance G_1 across the gap of length L in the on-state, can be written as follows [2]:

$$G_1 = \frac{4n_r}{(n_r + 1)^2} \frac{e}{(h\nu)} (\mu_n + \mu_p) \frac{E_1}{L^2} \qquad (6.1)$$

where n_r is the refractive index of the substrate, $h\nu/e$ is the photon energy in volts, $\mu_n + \mu_p$ is the sum of the electron and hole mobilities, E_1 is the energy of the turn-on optical pulse, which is incident on the gap. Typical values of the above parameters for the experimental switch are, n_r equal to 3.6, $h\nu/e$ equal to 2.34 V, $\mu_n + \mu_p$ equal to 2000 cm^2 V^{-1} s^{-1}, L equal to 0.034 cm, and E_1 equal to 10 μJ. G_1 computed by using (6.1) is 5.0 mhos (or $R_1 = 0.20\ \Omega$). This value is very small compared to the 50 Ω characteristic impedance of the microstrip, and hence is considered to be a good approximation of a series short circuit.

6.2.3 Gap Shunt Conductance

An expression for the gap shunt conductance G_2 in the off-state is presented in this section. The assumption made here is that the absorption depth is large compared to the substrate thickness D. Hence, multiple optical reflections between the substrate surfaces have to be taken into consideration. Summing these and integrating over the substrate thickness, the conductance G_2 connecting upper and lower conductor in the off-state is [2]:

$$G_2 = \frac{2\alpha_2}{D} \frac{e}{(h\nu)} (\mu_n + \mu_p) E_2 \qquad (6.2)$$

where α_2 is the absorption constant at the turn-off wavelength, D is the substrate thickness, and E_2 is the energy of the turn-off optical pulse. Typical values of the above parameters for the experimental switch are, $h\nu/e$ equal to 1.17 V, D equal to 0.043 cm, α_2 equal to 7 cm^{-1}, and E_2 equal to 10 μJ. G_2 computed using equation (6.2) is 5.6 mhos, or ($R = 0.18\ \Omega$). This value is very large compared to the 0.02 mhos characteristic admittance of the microstrip, and hence is considered to be a good approximation to a shunt short circuit.

6.2.4 Insertion Loss in the On-State

The finite gap resistance in the on-state gives rise to additional loss in the transmitted microwave signal and is expressed in terms of G_1 and Z_0 as follows [2]:

$$\text{Insertion Loss} = 10 \log_{10} \left| \frac{2 G_1 Z_0}{1 + 2 G_1 Z_0} \right|^2 \text{ dB} \qquad (6.3)$$

On substituting for G_1 from (6.1) and Z_0 equal to 50 Ω, the insertion loss of the switch in the on-state results as 0.02 dB.

6.2.5 Repetition Rate

The recombination of the optically-generated electrons and holes in silicon is rather slow. It is typically on the order of 1 μs. Hence, at least 1 μs must elapse before the switch can be reused. This imposes an upper limit on the repetition rate for the on-off cycle. Fortunately, there are several techniques that can be used to increase the recombination time in silicon. One such technique is to dope the silicon with gold atoms, which reduces the carrier lifetime to a few nanoseconds [5]. Alternatively, we can use a different substrate material such as GaAs which has a recombination time on the order of 100 ps [6]. A small recombination time implies that the switch requires a turn-on optical pulse only, and it would turn off automatically. Thus, with GaAs as the substrate material, repetition rate as high as 1 GHz should be possible, which would be more than adequate for most communication and radar applications.

6.2.6 Isolation in the Off-State

The isolation in the off-state is expressed as [2]:

$$\text{Isolation} = 10 \log_{10} \left| \frac{1 + 2 G_1 Z_0 + G_2 Z_0 + G_1 G_2 Z_0^2}{1 + 2 G_1 Z_0} \right|^2 \text{ dB} \qquad (6.4)$$

On substituting for G_1, G_2, and Z_0, the isolation of the switch in the off-state results as 43 dB. In deriving the above equation, the gap coupling capacitance C_c is neglected. The capaciative reactance due to C_c decreases as the frequency increases, thereby increasing the coupling and lowering the isolation. The gap coupling capacitance for the geometry under consideration as estimated from [4] is about 1.5 $\times$ 10^{-14}F.

In the case of substrate materials with picosecond recombination time, such as GaAs, the isolation of the switch in the off-state is a function of the conductance

G_g of the gap in the absence of any optical illumination, and the capacitance C_c. Hence, (6.4) gets modified as follows [2]:

$$\text{Isolation} = 10 \log_{10}\left\{\left|\frac{2 G_1 Z_0}{1 + 2 G_1 Z_0}\right|^2 \left[\frac{(1 + 2 Z_0 G_g)^2 + (2\omega_1 Z_0 C_c)^2}{4[(G_g Z_0)^2 + (\omega_1 Z_0 C_c)^2]}\right]\right\} \text{ dB} \quad (6.5)$$

where ω_1 is equal to $2\pi f$ and f is the frequency of operation. As an example, if we assume f to be equal to 10 GHz and G_g as 10^{-6} mhos, the computed isolation is 20.5 dB.

6.2.7 Switching Speed

The switching speed of the optoelectronic gate is limited by the combination of the following three factors: the time constant as determined from the product of the gap resistance and capacitance, the dielectric relaxation time, and the amplitude and duration of the optical pulse. The effects of the gap time constant and the dielectric relaxation time can be neglected as they are on the order of 1 ps or less [2]. However, the amplitude and duration of the optical illumination influence the rise and fall time, insofar as the photoconductivity increases linearly with the optical pulse energy.

6.2.8 Experimental Switch Performance and Discussions

In the experimental switch, the wavelength used to turn on the switch is chosen as 0.53 μm because it is easily generated by using an Nd:glass laser. The wavelength used to turn off the switch is chosen as 1.06 μm. This wavelength is obtained by generating the second harmonic of the 0.53 μm pulse to avoid the use of two lasers to control the switch.

The measured insertion loss of the switch at 1.0 GHz in the on-state is about 0.25 dB and is considered to be small. The measured isolation in the off-state is about 13 dB. The discrepancy between the isolation predicted by (6.4) and the measured value is because the analysis neglects the gap coupling capacitance, C_c. Thus, the upper frequency limit of operation for the switch is decided by the amount of isolation that can be achieved in the off-state.

The maximum CW RF power that can be switched without breakdown or excess heating is about 200 W. The peak RF power that the switch can handle is about 1 kW.

The measured rise and fall time of the switch is approximately 10 ps [2,3]. The width of the 0.53 μm and 1.06 μm optical pulses are 5 ps and 8 ps, respectively. Further, the optical energy in each of these pulses is about 10 μJ.

Finally, as has been experimentally observed, the gap dimensions do not degrade, even after several hundreds of on-off cycles, and hence the switch is considered to be reliable.

6.3 OPTOELECTRONIC SWITCHING AND ELECTRONIC SWITCHING PERFORMANCE COMPARISON

The performance of the optoelectronic switch is compared with conventional electronic switches which use GaAs MESFET and PIN diodes. The isolation of the GaAs MESFET and PIN diode switches are better than 25 dB at X-band and below [7,8]. To achieve so high an isolation, these switches use complex impedance-matching and tuning circuits [9]. The tuning circuits are also responsible for narrowing the bandwidth of the switch. However, the isolation of the optoelectronic switch is not very large, but it has the advantage of being simple to construct.

The power-handling capability and the switching speed of the electronic switches are interrelated. For example, switching time of GaAs MESFET and PIN diode switches are about 150 ps and 1 ns, respectively [7], when the RF power is in the mW range. For these switches to handle high RF power levels, the geometry of the devices involved must be made large. The large dimensions result in higher gate-to-source capacitance, C_{gs}, in the case of MESFETs and junction capacitance, C_j, in the case of PIN diodes. In addition, there is also the problem of minority-carrier storage effects in the case of PIN diodes. These additional capacitances lower the switching time. The power-handling capability and switching speed are extremely good in the case of optoelectronic switches as they do not suffer from the above disadvantages.

The driver circuit power expended in turning on a MESFET switch in 150 ps at a switching frequency of 5 MHz is typically about 114 mW [7]. The microstrip optoelectronic switch, however, requires a total of 20 μJ of optical energy to be incident on the microstrip gap for each on-off cycle. If we assume that this cycle can take place at a rate of 1 MHz, then the optical drive power that is expended is 20 mW. The typical overall efficiency of an Nd/glass laser is about 1% [10]; hence, the electrical power expended is about 2 W.

The MESFET as well as the PIN diode switches require special low-capacitance bias circuits with charging time significantly smaller than the switching time. However, the optoelectronic switches do not require any bias circuits.

Finally, the optoelectronic switch can handle moderately high microwave power, unlike GaAs MESFETs or PIN diodes, which are limited by the breakdown voltage and thermal dissipation. As a concluding remark, we can mention that, in a manner similar to electronic switches, the optoelectronic switching concept can be extended to the development of SPNT switches. Optoelectronic switches also can be realized using other planar transmission media such as coplanar waveguide [11] and integrated finline [12]. Also, monolithic integration with other microelectronic devices and circuit components [13] appears to be promising.

REFERENCES

1. Sze, S.M., *Physics of Semiconductor Devices,* 2nd Ed., New York, John Wiley and Sons, 1981, p. 750.
2. Johnson, A.M., and D.H. Auston, "Microwave Switching by Picosecond Photoconductivity," *IEEE J. Quantum Electronics,* Vol. QE-11, No. 6, June 1975, pp. 283–287.
3. Auston, D.H., "Picosecond Opto-electronic Switching and Gating in Silicon," *Applied Physics Letters,* Vol. 26, No. 3, February 1975, pp. 101–103.
4. Rahmat-Samii, V., T. Itoh, and R. Mittra, "A Spectral Domain Analysis for Solving Microstrip Discontinuity Problems," *IEEE Trans. Microwave Theory Tech.,* Vol. MTT-22, April 1974, pp. 372–378.
5. Ballis, W.M., "Properties of Gold in Silicon," *Solid-State Electronics,* Vol. 9, 1966, pp. 143–168.
6. Lee, C.H., "Picosecond Optoelectronic Switching in GaAs," *Applied Physics Letters,* Vol. 30, No. 2, January 1977, pp. 84–86.
7. Gaspari, R.A., and H.H. Yee, "Microwave GaAs FET Switching," *IEEE Int. Microwave Symp. Digest,* 1978, pp. 58–60.
8. Ayasli, Y., "Microwave Switching with GaAs FETS," *Microwave Journal,* Vol. 25, No. 11, November 1982, pp. 61–74.
9. White, J.F., *Microwave Semiconductor Engineering,* Van Nostrand Reinhold, New York, 1982, Chapter III.
10. Hecht, J., *The Laser Guidebook,* McGraw-Hill Book Company, New York, 1986, p. 292.
11. Castagne, R., S. Laval and R. Laval, "Picosecond 1-Wavelength Optoelectronic Gate," *Electronics Letters,* Vol. 12, No. 17, August 1976, pp. 438–439.
12. Uhde, K., "Optoelectronic Millimeter-Wave Switching Using a Finline-on-Silicon Substrate," *Electronics Letters,* Vol. 23, No. 21, October 1987, pp. 1155–1156.
13. Smith, P. R., D. H. Auston, and M. C. Nuss, "Subpicosecond Photoconducting Dipole Antenna," *IEEE J. Quantum Electronics,* Vol. 24, No. 2, February 1988, pp. 255–260.

Chapter 7
Optoelectronic Microwave Signal Generation

7.1 INTRODUCTION

Future active phased array antenna systems based on gallium arsenide microwave monolithic integrated circuit technology will have a transmitting module and a receiving module integrated with each element or small group of radiating elements. These MMIC circuits will require a reference microwave signal for synchronization of the transmitting module as well as the local oscillator in the receiving module. The use of conventional microwave transmission lines such as coaxial cables and rectangular metal waveguides for distributing the reference signal from a central unit would make the array prohibitively large, heavy, and complex. Also, coaxial cables suffer from excessive attenuation as the frequency increases. For example, a semi-rigid coaxial cable, 0.141 inch in diameter, will attenuate a 10 GHz signal at the rate of 50 dB/100 feet. As the frequency is increased to 100 GHz the attenuation reaches 500 dB/100 feet. Rectangular metal waveguides, although they have much lower attenuation, can impose physical restraints. As a result, there is a need to explore alternative techniques for the generation and distribution of microwave signals in future active phased arrays.

Single-mode optical fibers with attenuation of less than 1 dB/km [1], semiconductor laser diodes capable of being directly modulated at microwave frequencies as high as 10 GHz, and photodiodes able to detect an intensity-modulated signal as high as 15 GHz are commercially available [2]. Hence, a possible alternative is to use optical fiber and semiconductor laser diode technologies for the generation and distribution of microwave signals. This technique also has the advantage of greatly reducing the size and weight of the system. Figure 7.1 conceptually illustrates this technique.

This chapter presents an optoelectronic technique to generate a microwave signal. In this technique, the two simultaneously injection-locked modes of a semiconductor laser diode are mixed in a photodiode to generate the microwave signal. The

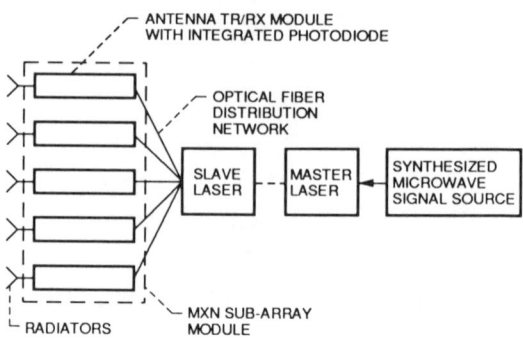

Figure 7.1 Schematic of a future MMIC based active phased array antenna system with optoelectronic microwave signal generation and distribution.

frequency range, temperature stability, power output and efficiency, pulling figure, effect of stray magnetic field, and mechanical shock and vibration are discussed and wherever possible compared with conventional electronic techniques of generating a microwave signal. In addition, because noise characteristics are often of critical importance in solid-state applications such as phased arrays, the noise performance of optoelectronic signal generators is discussed in detail in this chapter, and is shown to be undoubtedly superior when compared to electronic oscillators.

7.2 LONGITUDINAL MODE POWER SPECTRA OF A FREE-RUNNING LASER DIODE

Let us consider a laser oscillator having a cavity length far greater than the emission wavelength. However, the transverse dimensions are chosen so that only the lowest order transverse mode can propagate. Thus, the active region of this laser diode forms a rectangular resonant cavity. This cavity can support several longitudinal modes of oscillation. These modes each have a unique resonance frequency and wavelength. These resonance frequencies are influenced by the dielectric properties of the cavity boundaries, the gain profile within the cavity, the presence of free carriers that lower the refractive index, and the local increase in temperature that increases the refractive index. Neglecting these effects, the condition for resonance is that the cavity walls ought to be separated by an integral number of half-wavelengths. Thus, the condition for longitudinal resonance is

$$\frac{k\lambda_k}{2} = nL \tag{7.1}$$

Where k is the longitudinal mode number, λ_k is the free space wavelength of the radiation of the kth mode, n is the refractive index of the semiconductor material at this wavelength, and L is the axial length of the laser cavity. Substituting $f_k = c/\lambda_k$ gives the frequency of the kth mode as

$$f_\kappa = kc/2nL \qquad (7.2)$$

As the cavity is of several wavelengths, it can support several modes. The frequency separation between any two adjacent modes is

$$\Delta f = f_\kappa - f_{\kappa-1} = c/2nL \qquad (7.3)$$

The optical power spectra of this laser diode under free-running conditions would resemble Figure 7.2(a). As an example, consider a GaAs laser diode with L equal to 1.2 mm and operating at a wavelength of 0.84 μm. For GaAs, n is about 3.56 at the above wavelength. Hence, the frequency separation (Δf) between adjacent modes as computed from (7.3) is 35.1 GHz.

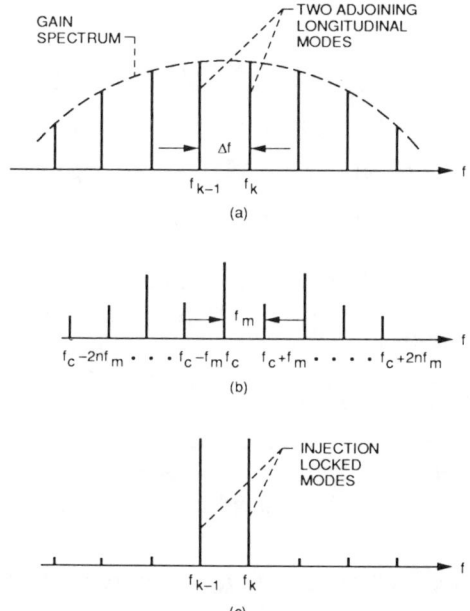

Figure 7.2 (a) Resonant power spectrum of the longitudinal modes of a long cavity laser diode; (b) power spectra of a directly frequency-modulated laser diode; (c) power spectra of the slave laser diode when injection locked.

7.3 POWER SPECTRA OF A DIRECTLY FREQUENCY-MODULATED LASER DIODE

The light output from a semiconductor laser diode varies with the change of injection current. Hence, a laser diode can be directly frequency-modulated (FM) by superimposing a sinusoidal signal at a frequency f_m on the dc bias of the laser diode. The electric field of the frequency-modulated optical wave can be mathematically represented as [3]:

$$E = E_0 \exp\{j[2\pi f_c t + \beta \sin(2\pi f_m t)]\} \tag{7.4}$$

where

f_c = laser center frequency of oscillation,
Δf = maximum frequency deviation,
f_m = modulation frequency,
β = frequency modulation index equal to $\Delta f/f_m$.

Furthermore, (7.4) can be expanded in terms of Bessel functions of the first kind as [3]:

$$\begin{aligned} E = &\ J_0(\beta)\, E_0 \sin(2\pi f_c t) + J_1(\beta)\, E_0 \sin\{2\pi(f_c + f_m)t\} \\ &- J_1(\beta)\, E_0 \sin\{2\pi(f_c - f_m)t\} + \ldots + J_n(\beta)\, E_0 \\ &\cdot \sin\{2\pi(f_c + n f_m)t\} + \ldots - J_n(\beta)\, E_0 \\ &\cdot \sin\{2\pi(f_c - n f_m)t\} \end{aligned} \tag{7.5}$$

Thus, the resulting spectrum of the laser has several sidebands space by f_m on either side of the optical carrier frequency, f_c. These sidebands are denoted as the frequency-modulated sidebands of the laser diode. The intensity of these sideband components are given by the square of the coefficient of the corresponding sideband terms in equation (7.5). Figure 7.2(b) shows a typical power spectrum of a directly frequency-modulated laser diode. As an example, if f_m is 5.846 GHz, the frequency separation between the $n = -1$ and $n = +1$ sideband is 11.692 GHz, whereas that between the $n = -2$ and $n = +2$ sideband is 23.384 GHz, and so forth.

7.4 FM SIDEBAND INJECTION LOCKING TECHNIQUE

There are several techniques that are used to injection-lock a laser diode. However, the FM sideband injection-locking technique that is discussed here involves the use of two separate laser diodes. One of these laser diodes has a long cavity, and hence can oscillate with several longitudinal modes as explained in Section 7.2. This laser is designated as the "slave laser." Let us focus our attention on any two adjacent longitudinal modes of this laser, and designate their optical frequencies as $f_{\kappa-1}$ and f_κ. Let the frequency separation between these modes be denoted as Δf.

The second laser is designated as the "master laser." The master laser is directly modulated by a sinusoidal signal at a frequency f_m. As a result, the FM sidebands formed are located at points $f_c \pm nf_m$ along the frequency axis, as discussed in Section 7.3.

The objective here is to simultaneously injection-lock the two adjoining longitudinal modes of the slave laser by the FM sidebands of the master laser. To do this, the optical emission from the master is coupled to the slave cavity. Further, the frequency f_m is adjusted such that $2nf_m$ is equal to Δf. The slave is then thermally tuned to translate the spectrum until $f_c + nf_m = f_\kappa$ and $f_c - nf_m = f_{\kappa-1}$ and injection-locking occurs with the total output power of the slave laser being transferred to the two injected modes. Figure 7.2(c) shows the power spectrum of the slave laser when injection-locked.

There are two possible mechanisms simultaneously occurring inside the laser diode when it is being injection-locked. The first mechanism causes the injected field to add an out-of-phase component, thereby causing a change in the resonance frequency of the slave laser field. The second mechanism involves a change in the gain required to maintain steady-state slave laser intensity. The change in gain is also accompanied by a shift in the resonance frequency.

7.5 MICROWAVE SIGNAL GENERATION TECHNIQUE

The simultaneously injection-locked longitudinal modes of the slave laser are used to generate the microwave signal. This is done by coupling the optical emission from the slave laser to a photodiode. The requirement for the photodiode is to be capable of responding to the wavelength of the slave laser emission. Also, the -3 dB roll-off frequency of the photodiode should be well above the signal frequency to be generated. The nonlinearity of the photodiode voltage-current characteristic generates a "beat signal" between the two slave laser modes. If P_1 and P_2 are the optical power in the two longitudinal modes, the generated microwave power P_m can be expressed as [4]:

$$P_m = 2R^2 \eta \, P_1 \, P_2 \, R_L \tag{7.6}$$

where

R = responsivity of the photodiode,
R_L = load resistance,
η = coupling efficiency of the light into the photodiode.

As an example, if P_1 and P_2 are each equal to 100 μW, R is 0.45 A/W, η is 0.5, and R_L is 50 Ω, then P_m is -40.0 dBm.

7.6 EXPERIMENTAL GENERATOR PERFORMANCE AND DISCUSSIONS

The experimental setup is shown in Figure 7.3(a). The slave laser has a cavity length of 1.2 mm, and consequently the frequency separation between the longitudinal modes is 35.1 GHz. The master laser is directly frequency-modulated at 5.846 GHz. Hence, the $n = +3$ and $n = -3$ FM sidebands of the master laser are used to injection-lock the slave laser. Simultaneous injection-locking of the adjoining slave laser longitudinal modes is achieved by adjusting the slave laser temperature, until the center frequencies of the modes match the FM sidebands. The two injection-locked longitudinal modes beat together in a photodiode thereby generating a 35 GHz signal. Figure 7.3(b) shows the microwave spectrum of the beat signal which has a 3 dB spectral width of 6 Hz. A practice in electrical engineering is to specify the spectral

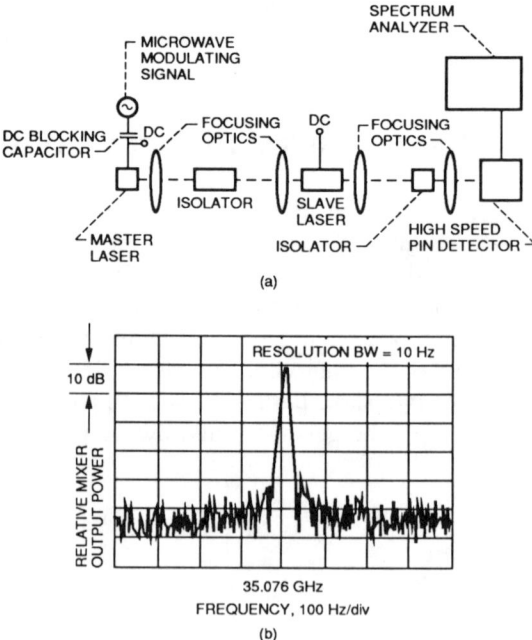

Figure 7.3 (a) Simplified experimental setup; (b) microwave spectrum of the generated signal.
Source: Goldberg, L., A.M. Yurek, H.F. Taylor, and J.F. Weller, "35 GHz microwave signal generation with an injection-locked laser diode," *Electronics Letters*, Vol. 21, No. 18, August 1985, pp. 814–815. Reprinted with permission.

purity of the signal in terms of the close-in carrier-to-noise ratio or the FM noise, instead of the 3 dB spectral width. This aspect will be discussed in greater detail in the next section. The injection-locking bandwidth of the slave laser is about 350 MHz.

7.7 CAPABILITIES AND LIMITATIONS

7.7.1 Frequency Range

The frequency limit of operation of this technique is governed by three factors. The first limitation arises from the length of the slave laser cavity. This length determines the frequency difference between the two longitudinal modes, which in turn decide the frequency of the generated microwave signal. A laser cavity length in the range of 250 to 1200 μm is typical. This translates to a frequency range of 35 to 170 GHz according to (7.3). The second limitation stems from the relaxation oscillation frequency of the master laser, the frequency up to which the master laser can be directly modulated. Beyond this frequency, the frequency response of the laser drops at the rate of -40 dB/decade. Commercially available GaAlAs lasers are capable of operation up to 10 GHz [2]. However, laboratory experimental laser diodes with relaxation frequencies as high as 30 GHz have been demonstrated [5]. The third limitation is due to the bandwidth of the photodiode. Commercially available photodiodes are capable of direct detection up to 15 GHz [2], while laboratory experimental diodes with -3 dB cut-off frequency as high as 100 GHz have been demonstrated [6].

Commercially available Gunn and IMPATT diodes are capable of operating at frequencies as high as 95 GHz and 145 GHz, respectively [15]. However, because of the finite response time, the estimated upper frequency limit is 150 GHz and 300 GHz in the case of the Gunn and IMPATT diodes, respectively [16].

7.7.2 Signal-to-Noise Ratio

Because the optical power at the photodiode is quite high, laser noise dominates over the photodiode or receiver noise. Hence, the noise floor of the optoelectronic microwave signal generator is decided by the intrinsic relative intensity noise (RIN) of the laser. The RIN arises from the shot noise processes associated with carrier injection and recombination inside the laser active layer. The disadvantage of RIN is manifest when the laser is directly modulated by a sinusoidal microwave signal. The RIN mixes with the modulation signal and generates noise sidebands on the carrier. These noise sidebands are observed to be strikingly similar to the sidebands of a frequency-modulated carrier signal. For this reason, the RIN is also known as the

FM noise or the *phase noise*. For a typical laser diode, RIN varies between -125 and -140 dBc/Hz [7]. Experiments with commercially available laser diodes have confirmed this and the best close-in signal-to-noise ratio (S/N) at 10 kHz, offset frequency relative to a 10 GHz modulation signal, is on the order of -130 dBc/Hz [8]. The RIN is expected to reduce further as laser diodes with higher relaxation frequencies become available.

The effect of the RIN on the local oscillator (LO) performance in a receiver module can be understood by considering Figure 7.4. In a communication system, let f_1 and f_2 represent two closely spaced RF signal frequencies that are to be down-converted. The RIN increases the noise power at the mixer output at frequencies close to the signal. This effectively degrades the receiver sensitivity to a nearby weak signal, such as f_2. Thus, one effect of local oscillator noise is to cause receiver desensitization due to a strong signal. Doppler radar systems use the frequency shift of the return echo to determine the velocity of the target. Local oscillator phase noise causes the large echo from stationary objects such as the ground, buildings, and other structures to mask the desired target signal, thus limiting detection sensitivity. Furthermore, phase noise spreads the target return energy, thereby degrading system range resolution.

The typical close-in S/N or the FM noise of electronic oscillators fabricated with a GaAs MESFET, silicon bipolar transistor, GaAs/GaAlAs hetrojunction bipolar transistor, silicon IMPATT diode, and InP Gunn diode are compared in Table 7.1. The S/N ratios quoted here are about the best that can be achieved with a particular oscillator. The low-frequency noise in these devices arises from traps within the active region, the semiconductor surface, or at the interface between the active region and the buffer layer or substrate. Consequently, GaAs devices are observed to have a higher low-frequency noise than silicon devices. Comparing the two techniques, we may observe that the optoelectronic technique undoubtedly has superior close-in S/N. The amplitude modulation (AM) noise is not of much concern as it is smaller than -130 dBc/Hz.

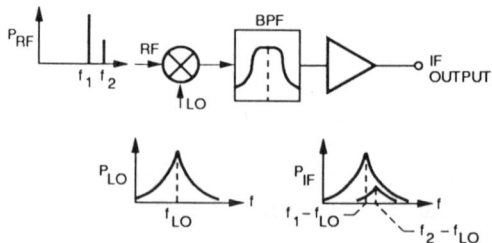

Figure 7.4 Effect of local oscillator noise on receiver sensitivity.

Table 7.1
Signal-to-Noise Ratio of Microwave Semiconductor Transistor and Diode Oscillators at 10 kHz Offset Frequency

Device Type	Frequency	Power (dBm)	Phase Noise (dBc/Hz)	Stabilization Technique	Reference
GaAs MESFET	X-Band	—	−65	—	[11]
GaAs MESFET	X-Band	—	−95	Dielectric Resonator	[11]
GaAs MESFET	X-Band	—	−120	Frequency-Locked Loop	[11]
GaAs/GaAlAs Heterojunction Bipolar Transistor	C-Band	10.2	−95	Dielectric Resonator	[12]
Si Bipolar Transistor	C-Band	14.9	−108	Dielectric Resonator	[12]
Si IMPATT	W-Band	21.6	−68	Resonant Cap in Full Height Waveguide	[13]
InP Gunn	W-Band	13	−60	TE_{011} Mode Cylindrical Cavity	[14]

7.7.3 Temperature Stability

The required temperature stability can be estimated from the linear thermal expansion coefficient α_L of the GaAs, which is $0.6 \cdot 10^{-5}/°C$ and the thermal refractive index coefficient α_n of GaAs, which is $0.99 \cdot 10^{-4}/°C$ [3]. Based on these values, the increase in the cavity length and the refractive index of the laser in the example of Section 7.2 is $7.2 \cdot 10^{-6}$ mm and $3.5244 \cdot 10^{-4}$, respectively. The resulting decrease in the Δf as computed from (7.3) is 3.686431 MHz/°C. Let us compare this value with that of a typical metal cavity-stabilized Gunn oscillator.

A typical oscillator of this type has a temperature coefficient of about −28 ppm/°C, which, at 10 GHz, results in −280 kHz/°C change in the frequency [9]. This is one order of magnitude smaller than that of the laser. Hence, the laser diode requires more precise control of the ambient temperature.

7.7.4 Power Output and Efficiency

Gunn and IMPATT diode oscillators are capable of generating several hundreds of mW of CW microwave power. The efficiency of GaAs Gunn diodes can vary from 6 to 2% over the frequency range of 35 to 95 GHz. The efficiency of silicon IMPATT diodes can vary from 13.5 to 5% over the frequency range of 35 to 140 GHz [15].

However, the power of the optoelectronically generated microwave signal is very small, typically in the range of -40 to -50 dBm. However, this does not pose a serious problem because MMIC receiving modules in phased arrays invariably have a low-noise amplifier (LNA) cascaded with a gain control amplifier, and the transmitting modules have a variable-gain-control (VGC) amplifier cascaded with a high-power amplifier stage to boost the signal to the desired level. The LNAs typically have a gain of about 12 dB and a noise figure of 5 dB at 30 GHz. The VGC amplifier has a dynamic range of about 30 dB [10]. The laser device efficiency is defined as

$$\eta_d = P/VI \qquad (7.7)$$

where P is the laser output power and, V and I are the respective dc voltage and current supplied to the device. For a GaAlAs laser diode, the efficiency can be as high as 20% [17]. A practice is to express the efficiency of photodiodes in terms of their quantum efficiency, the number of electron-hole pairs generated per incident photon. The quantum efficiency for a GaAlAs-GaAs photodiode is as high as 65% [2].

7.7.5 Pulling Figure

The load *voltage standing wave ratio* (VSWR) affects the frequency of oscillation of electronic oscillators. A measure of how easily the oscillator is detuned by a reactive load is the pulling figure. The pulling figure is given by the expression [9]

$$f = \frac{f_o}{Q_{ext}}\left[r - \frac{1}{r}\right] \qquad (7.8)$$

where

f = is the change in the oscillator center frequency
f_o = oscillator center frequency
Q_{ext} = external Q of the oscillator cavity
r = load VSWR

As an example, if the external Q of the cavity is 1000, the return loss is -12 dB, and the center frequency is 10 GHz, then the pulling figure computed from (7.8) is 10.71 MHz. Optoelectronic signal sources are unaffected by the load VSWR, insofar as there is an inherent isolation.

7.7.6 Stray Magnetic Field

There is evidence to show that the oscillation frequency of GaAlAs laser diodes shifts under the influence of an external magnetic field. Typically, this shift at room temperature is about 500 MHz at a magnetic flux density of 1.4 Wb/m^2 [18]. Hence, optoelectronic signal generators may require shielding from stray magnetic fields.

7.7.7 Shock and Vibrations

Laboratory experiments have shown that bulk as well as surface acoustic waves in the power and frequency ranges of a few mW and kHz, respectively, are capable of modifying both the resonant cavity length and refractive index of the material of a semiconductor laser diode. These changes have negligible effects on the intensity of the laser output, but cause frequency modulation of the optical carrier frequency [19]. This suggests that mechanical shock and vibrations may affect the performance of optoelectronic signal generators.

REFERENCES

1. Corning SMF-21 Optical Fiber Product Specification Sheet, Corning Glass Works, Corning, NY, February 1987.
2. Browne, J., "Window Boosts Laser Diode Power," Ortel Corp. Product Technology, *Microwaves & RF*, Vol. 24, June 1985, pp. 159–160.
3. Kobayashi, S., Y. Yamamoto, M. Ito, and T. Kimura, "Direct Frequency Modulation in AlGaAs Semiconductor Laser," *IEEE J. Quantum Electronics*, Vol. QE-18, No. 4, April 1982, pp. 582–595.
4. Goldberg, L., A.M. Yurek, H.F. Taylor, and J.F. Weller, "35 GHz Microwave Signal Generation with an Injection-locked Laser Diode," *Electronics Letters*, Vol. 21, No. 18, August 1985, pp. 814–815.
5. Uomi, K., T. Mishima, and N. Chinone, "Ultrahigh Relaxation Oscillation Frequency (up to 30 GHz) of Highly p-doped GaAs/GaAlAs Multiple Quantum Well Lasers," *Applied Physics Letters*, Vol. 51, No. 2, July 1987, pp. 78–80.
6. Wang, S.Y., and D.M. Bloom, "100 GHz Bandwidth Planar GaAs Schottky Photodiode," *Electronics Letters*, Vol. 19, No. 14, July 1983, pp. 554–555.
7. Lau, K.Y., and H. Blauvelt, "Effect of Low-Frequency Intensity Noise on High-Frequency Direct Modulation of Semiconductor Injection Lasers," *Applied Physics Letters*, Vol. 52, No. 9, February 1988, pp. 694–696.
8. Newberg, I.L., C.M. Gee, G.D. Thurmond, and H.W. Yen, "Radar Applications of X-Band Fiber Optic Links," *IEEE Int. Microwave Symp. Digest*, 1988, pp. 987–990.
9. Hamilton, S., "FM and AM Noise in Microwave Oscillators," *Microwave J.*, Vol. 21, No. 6, June 1978, pp. 105–109.
10. Geddes, J., V. Sokolov, D. Carlson, P. Bauhahn, and R.R. Romanofsky, "Characteristics of 30 GHz MMIC Receivers for Satellite Feed Array Applications," *IEEE GaAs Integrated Circuit Symp. Digest*, 1987, pp. 155–158.
11. Bianchini, M.J., J.B. Cole, R. DiBiase, Z. Galani, R.W. Laton, and R.C. Waterman, Jr., "A Single-Resonator GaAs FET Oscillator with Noise Degeneration," *IEEE Int. Microwave Symp. Digest*, 1984, pp. 270–273.
12. Agarwal, K.K., "Dielectric Resonator Oscillators Using GaAs/GaAlAs Heterojunction Bipolar Transistors," *IEEE Int. Microwave Symp. Digest*, 1986, pp. 95–98.

13. Brookbanks, D.M., A.M. Howard, and M.R.B. Jones, "Si IMPATTs Exhibit Low Noise at MM-Waves," *Microwaves & RF*, Vol. 22, No. 2, February 1983, pp. 68–72.
14. Smith, D.C., T.J. Simmons, and M.R.B. Jones, "A Comparison of the Performance of Millimeter-Wave Semiconductor Oscillator Devices and Circuits," *IEEE Int. Microwave Symp. Digest*, 1983, pp. 127–129.
15. Millimeter-Wave Products Catalog, Hughes Aircraft Company, pp. 42, 78, 79.
16. Sze, S.M., *Physics of Semiconductor Devices*, 2nd Ed., New York, John Wiley and Sons, 1981, pp. 597, 671.
17. *Laser Diode Product Catalog*, Spectra Diode Laboratories.
18. Sato, T., S. Yashima, and M. Shimba, "Frequency Shift of a GaAlAs Diode Laser in a Magnetic Field," *Electronics Letters*, Vol. 22, No. 19, September 1986, pp. 979–981.
19. Greenhalgh, P.A., and P.A. Davies, "Direct Frequency Modulation of a Semiconductor Laser by Surface Acoustic Waves," *Electronics Letters*, Vol. 21, No. 14, July 1985, pp. 598–599.

Chapter 8
Optoelectronic Switch Matrix

8.1 INTRODUCTION

The information-handling capability of a communication satellite is vastly increased by the reuse of the uplink and downlink frequencies. One technique of achieving frequency reuse is to use an antenna aboard the satellite, which is capable of generating several shaped beams, each operating at the same frequency, but directed to a different region on the continent. In such a system, the transfer of information from one region to another is made possible by reconfiguring the interconnectivity among the uplink and the downlink beams. In practice, this interconnectivity is accomplished by a switch matrix on board the satellite. The switch matrix could operate either with conventional microwave switches based on MESFET and PIN diodes, or with optoelectronic switches based on *avalanche photodiodes* (APD) or PIN photodiodes.

The optoelectronic switch matrix has several advantages over conventional semiconductor switch matrices. First, the distribution of the input signals to the crosspoints is done by an optical power divider. This ensures that there is no crosstalk between the input distribution lines. Second, as photons and electrons do not interact, there is no leakage of the input optical signals to the output electrical lines. This ensures high isolation between the input and output lines. Third, the signal reflected back into the power divider is independent of the state of the switches, as the optical reflectance of a photodetector is unaffected by its bias. This eliminates the need for RF impedance-matching networks. Finally, the semiconductor laser diode, optical power divider, and optical detectors have the potential to be monolithically integrated on a single substrate, which can result in enhanced performance, greater reliability, lighter weight, and reduced volume. The main disadvantage of optoelectronic switching is that, if avalanche photodiodes are used, voltages on the order of 100 to 150 V must be switched within nanoseconds, resulting in large electromagnetic interference (EMI).

Section 8.2 explains the principle of the optoelectronic switch matrix. Section 8.3 describes the principle of a silicon APD crosspoint switch. Section 8.4 presents experimental optoelectronic switch performance. Section 8.5 compares the performance of the optoelectronic switch matrix and electronic switch matrix.

8.2 PRINCIPLE OF AN OPTOELECTRONIC SWITCH MATRIX

An optoelectronic switch matrix [1] has input and output ports that are purely electrical networks. However, internally, an incoming electrical signal is made to intensity-modulate a semiconductor laser diode. The intensity-modulated light is then distributed to a row of photodiodes in a crosspoint matrix. The distribution is accomplished by optical fiber or waveguide components such as power dividers. An alternative scheme is to allow freely propagating light to fall upon a lens array that focuses a portion of the source power on each of the photodiodes.

The photodiodes at the crosspoint are connected in series with a load resistance and a bias network. In addition, the load resistor is coupled through an amplifier to one of the electrical output lines to which we may desire to transfer the input signal. The sensitivity of the photodiode to an optical signal is enhanced by reverse biasing the photodiode, thus creating a low insertion loss path between the electrical input line and the output line connected to the load resistor. However, the sensitivity of the photodiode to an optical signal is reduced by forward biasing the photodiode, yielding a large isolation between the input and the output lines. Clearly, by extending this principle to a row of photodiodes in a crosspoint matrix, a signal path can be established at will between the incoming line and any of the outgoing lines. The arrangement described above constitutes a 1-by-N crosspoint switch, where N is the number of photodiodes employed. An M-by-N crosspoint matrix can be constructed by connecting M such devices electrically in parallel, as shown schematically in Figure 8.1, where $M = N = 3$.

8.3 AVANLANCHE PHOTODIODE OPTOELECTRONIC CROSSPOINT SWITCH

8.3.1 Principle of Operation

The technique of optoelectronic switching with an avalanche photodiode exploits the enhancement in its response to an optical signal when the bias across the diode is switched from the zero bias state to the reverse bias condition, reverse bias being the normal condition for photodetection. Thus, at the crosspoint, the APD recovers the RF signal from the optical carrier and routes the RF signal to an assigned output port.

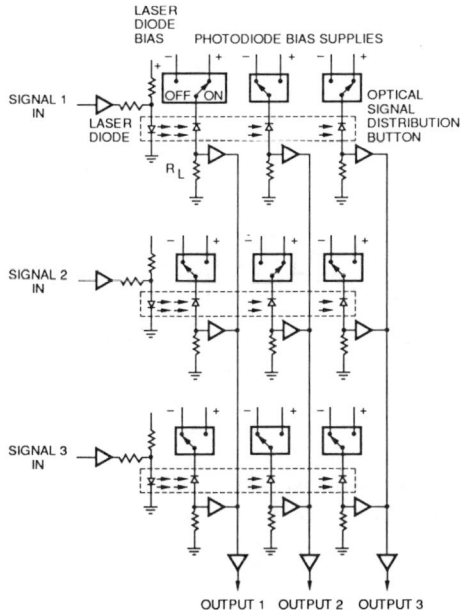

Figure 8.1 Schematic representation of an optoelectronic 3 × 3 matrix switch.
Source: MacDonald, R.I., and E.H. Hara, "Switching with Photodiodes," *IEEE Quantum Electronics,* Vol. QE-16, No. 3, March 1980, pp. 289–295. Reprinted with permission.

8.3.2 Physical Origin of the Equivalent Circuit Model Elements

The APD is modeled for the purpose of circuit analysis as shown in Figure 8.2(a). In this equivalent circuit model, the current source of magnitude I_{ph} and the *RC* network have their origin within the device as discussed below. Let us assume that the device is optically illuminated. When optical illumination is incident on the active area of the device, excess carriers are generated. The rate g at which these carriers are generated is given by

$$g = P\eta/h\nu \tag{8.1}$$

where

P = input optical signal power,
η = quantum efficiency,
h = Planck's constant,
ν = frequency of the optical signal.

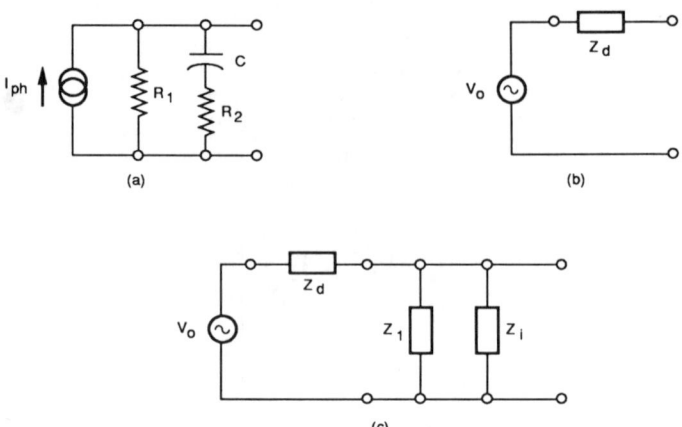

Figure 8.2 (a) Equivalent circuit model of the APD optoelectronic switch; (b) Thevenin equivalent circuit model; (c) equivalent circuit taking into consideration load impedance and external circuit impedance.

Further, if we assume that the APD is reverse biased, avalanche multiplication of the generated carriers occurs. The multiplication factor M is given by [3]:

$$M = \frac{1}{1 - (V_r/V_b)^n} \qquad (8.2)$$

where

V_r = reverse bias voltage,
V_b = breakdown voltage,
n = a constant.

As a numerical example, if V_r = 98 V, V_b = 100 V, and n = 2, then the multiplication factor M computed by using (8.2) is 25.3. As a consequence of avalanche multiplication, the number of net carriers N available for conduction is

$$N = M(P\eta/h\nu) \qquad (8.3)$$

The current I_{ph} due to these excess carriers is then

$$I_{ph} = eN \qquad (8.4)$$

where e = electronic charge. As a numerical example, if $P = 50$ μW, $\eta = 75\%$, and the wavelength of the optical illumination $\lambda = 0.85$ μm for the APD in the last example, the photocurrent I_{ph} computed from (8.4) is 0.65 mA.

The RC network represents the diode impedance under reverse bias condition. The diode impedance is dominated by the resistance and capacitance of the depletion region which is a consequence of the reverse bias. Typical values are $R_1 > 10^8$ Ω, $R_2 \ll 1$ Ω, and $C \approx 5$ pF [1], and these values are essentially independent of the reverse bias voltage.

In Figure 8.2(b), the equivalent circuit model of Figure 8.2(a) is replaced by its Thevenin equivalent circuit. In this figure, Z_d represents the lumped impedance of the diode, and V_0 is the Thevenin equivalent voltage source and is equal to

$$V_0 = I_{ph} Z_d \tag{8.5}$$

On substituting from equations (8.3) and (8.4) yields [2]:

$$V_0 = (Pe\eta/h\nu) M Z_d \tag{8.6}$$

8.3.3 Signal Transmission in the On-State

An APD almost always works in conjunction with a load resistance Z_1. In addition, the input impedance Z_i of an external circuit connected across Z_1 influences the APD performance. The resulting equivalent circuit model is as shown in Figure 8.2(c). Based on this model, the electrical signal power P_e delivered to an external circuit in response to the input optical signal can be written as

$$P_e = \frac{|Z_1 V_0|^2}{|Z_1 + Z_d|^2} \frac{Z_i}{|Z_i + Z_s|^2} \tag{8.7}$$

where Z_s is the source impedance formed by the parallel combination of Z_d and Z_1. Substituting for V_0 from (8.6) in (8.7) yields [2]:

$$P_e = \frac{|Z_1|^2}{|Z_1 + Z_d(r)|^2} \frac{Pe\eta(r)^2}{h\nu} M^2(r) Z_d(r) \tag{8.8}$$

where the subscript r denotes the APD parameters under the reverse bias condition. As a numerical example, if $Z_1 = 50$ Ω, $Z_d(r) = 1$ Ω, $P = 100$ μW, $\eta(r) = 75\%$, $\lambda = 1$ μm, and $M(r) = 100$, then, from (8.8), P_e is equal to 36.4 μW.

8.3.4 Isolation in the Off-State

The isolation I of the optoelectronic switch is defined as

$$I = P_e(r)/P_e(f) \tag{8.9}$$

where $P_e(r)$ and $P_e(f)$ are the electrical signal power delivered to the external circuit when the diode is reverse biased and when it is forward biased, respectively. Substituting for these quantities from (8.8), the following expression results [2]:

$$I = \frac{\eta(r)^2}{\eta(f)} \frac{M^2(r)}{M^2(f)} \frac{|Z_d(r)|^2}{|Z_d(f)|^2} \frac{|Z_i + Z_s(f)|^2}{|Z_i + Z_s(r)|^2} \tag{8.10}$$

where the subscripts r and f indicate appropriate quantities when the diode is reverse biased and forward biased, respectively. As a numerical example, if $\eta(r) = 75\%$, $\eta(f) = 5\%$, $M(r) = 80$, $M(f) = 1$, $Z_d(r) \approx Z_d(f)$ and $Z_s(r) \approx Z_s(f)$, then the isolation I computed by using (8.10) is 61.6 dB.

8.4 EXPERIMENTAL SWITCH PERFORMANCE AND DISCUSSIONS

In the experimental optoelectronic crosspoint switch [2], a silicon APD is used as the switching element. A GaAlAs semiconductor laser diode operating at a wavelength of 0.82 μm is used as the optical source. At this wavelength, the responsivity for the silicon APD is $0.65 \times M$ μA/μW, [3] which is considered sufficiently large. Further, the laser diode is directly frequency-modulated by superposing a microwave signal on the dc bias to the laser.

As described in Section 8.3.1, the switch is turned on and off by, respectively, reverse biasing and forward biasing the APD. The measured isolation is about 80 dB over the frequency range 10 MHz to 1 GHz, and degrades to 60 dB at 3 GHz, and further down to 50 dB at 4 GHz [2]. The high isolation is the result of the quantum efficiency reduction and the gain elimination with the change from reverse to forward biasing condition.

The measured turn-on time is about 400 ns. The slow turn-on time may be improved if the charge storage in the junction can be avoided. This is possible if the off-state can be established without applying a forward bias. Switching time on the order of 20 ns has been subsequently demonstrated by this technique [4].

The insertion loss in the on-state arises due to the process of converting electrons to photons and photons to electrons at the laser diode and the photodiode, respectively. In addition, losses occur at the couplings between the optical fiber and the laser diode at the transmitting end or the photodiode at the receiving end. The conversion at the laser diode is normally expressed in terms of the differential quantum efficiency, which for a typical GaAlAs laser diode varies between 25 and 50% [5]. The conversion at the photodiode is expressed in terms of the quantum efficiency which as assumed in Sections 8.3.2 and 8.3.3 for a typical silicon APD, is about 75% at the optical wavelength of the laser diode [6]. The coupling efficiency between the optical fiber and the devices can range between 50 to 80% [3]. Hence, taking into consideration the above factors, the overall insertion loss is about 20 dB. The insertion loss can be reduced if the APD is biased for an avalanche gain. Typically, the gain of an APD is about 15 to 20 dB [4]. Therefore, the insertion loss reduces to a negligible value. Both the insertion loss and the isolation have a tendency to degrade as the frequency increases. One reason for this is because of the roll-off in the laser diode and the photodiode frequency response. The respective parasitic circuit elements are responsible for limiting the bandwidth of these devices.

8.5 OPTOELECTRONIC SWITCH MATRIX AND ELECTRONIC SWITCH MATRIX PERFORMANCE COMPARISON

The obvious performance parameters such as insertion loss, isolation, and switching time, as well as a few salient features of the optoelectronic switch matrix are compared with those of the electronic switch matrix in Table 8.1. The electronic or microwave switch matrix technology is fairly mature. Several proof-of-concept models

Table 8.1
Optoelectronic Switch Matrix and Electronic Switch Matrix Performance Comparison

Description	Optoelectronic	Electronic
Switch Matrix Size	3 × 3	8 × 8
Configuration	Crossbar, Detector Switched	Crossbar
Frequency Band (GHz)	0.01–4.0	3.7–4.2
Switching Device	Silicon APD	GaAs Dual-Gate MESFET
Bias Voltage (V)	−125	4
Insertion Loss (dB)	5	8.4
Isolation (dB)	80 (1 GHz), 60 (3 GHz), 50 (4 GHz)	60
Switching Time (nS)	20	30
Group Delay (nS/path)	—	0.3
Size (cm)	—	15,700
Weight (kg)	—	13.98
Power Consumption (W)	10–100 mW/crosspoint	17.6
RF Power Dividers	No	Yes
RF Power Combiners	Yes	Yes
References	[1,2,4,22]	[23]

have been manufactured, and extensive electrical, environmental, and radiation testing has been done [7]. However, similar studies have not been made for an optoelectronic switch matrix, and hence a complete comparison is very difficult at present. However, data are available on the effects of vibration, temperature, and nuclear radiation damage for discrete optoelectronic devices. Some reliability statistics are also available in the literature. The effects of vibration and temperature on the laser diode performance have been discussed in Chapter 7. The intermodulation distortion, signal-to-noise ratio, radiation damage, and reliability aspects are discussed below.

8.5.1 Intermodulation Distortion

The nonlinearity of the laser diode drive current *versus* the optical power intensity characteristic can generate higher harmonic distortion signals as well as third-order IMD products. The *second harmonic distortion* (2HD), *third harmonic distortion* (3HD), as well as higher order harmonic distortion-generated signals are not of much concern, as they do not fall within the frequency band of interest. However, when two closely spaced signals at frequencies f_1 and f_2 within the frequency band of interest are made to modulate the laser diode, third-order IMD products can be generated at frequencies $2f_1-f_2$ and $2f_2-f_1$, which in most probability will lie within the bandwidth of the channel, and are thus undesirable. The amplitude of the IMD products are in general a function of the dc laser bias current, *the optical modulation depth* (OMD), the frequency of the modulating signal relative to the relaxation oscillation frequency, f_r, and the magnitude of the relaxation oscillation peak [8,9].

Experiments have demonstrated that, regardless of the constructional features of the laser diode, the level of distortion is the same for all laser diodes at a given f_r and OMD [9]. The following is a summary of the observed significant features [8,9]:

1. The relative amplitudes of the 2HD and 3HD have maxima at frequencies near $f_r/2$. Typical relative amplitude values are -8 and -18 dB for 2 HD and 3HD, respectively, at $f_r/2$, when OMD is 0.8.
2. The relative amplitude of the 2HD increases as the square of the OMD at a fixed f_r.
3. The relative amplitude of the IMD has maxima at frequencies near $f_r/2$ and f_r, but the level at f_r is higher. Typically, for OMD of 0.8, the relative amplitude of the IMD is -25 and -30 dB at f_r and $f_r/2$, respectively.
4. The relative amplitude of the IMD increases as the cube of the OMD at a fixed f_r.
5. The relative amplitude of the 2HD and IMD decreases rapidly below $f_r/2$.
6. Lasers with large resonance peaks in the small-signal frequency response also exhibit pronounced maxima in the distortion curves.

Hence, to keep the third-order IMD products below an acceptable value, the modulating frequency must be less than $f_r/2$ or a laser diode with a higher f_r must be chosen. Figure 8.3 schematically illustrates the interrelationship between the small-signal frequency response of the laser diode and the relative amplitude of the distortion signals.

The measured third-order IMD in the case of a 20 × 20 microwave switch matrix designed to operate over the frequency range of 3.5 to 6.0 GHz, varies from −28 to −42 dB [10].

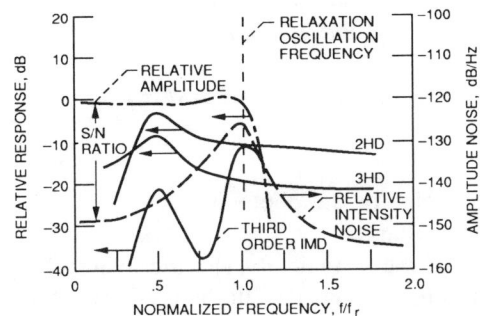

Figure 8.3 Interrelationship between the small-signal frequency response of the laser and the relative amplitude of the distortion signals.

8.5.2 Signal-to-Noise Ratio

Let us examine a single crosspoint optoelectronic switch, which is shown in Figure 8.4. Let us assume that an APD is used as the photodetector. Further, let us assume that the amplifier following the switch has an input impedance that is much greater than the load resistance R_1. This ensures that the amplifier thermal noise is much smaller than that of the load resistance R_1, and hence can be neglected. The load resistor R_1 contributes a mean square thermal noise or Johnson noise current [3]:

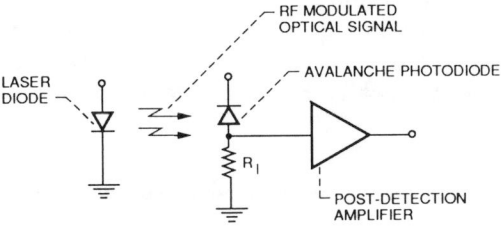

Figure 8.4 Isolated optoelectronic crosspoint switch and amplifier circuit.

$$\langle i_T^2 \rangle = 4\,kTB/R_1 \tag{8.11}$$

where

k = Boltzmann's constant,
T = absolute temperature,
B = bandwidth.

The other noise sources associated with the photodetector are the quantum noise and dark current noise. The quantum or shot noise arises from the statistical nature of the production and collection of photoelectrons when an optical signal is incident on the photodetector. The mean square quantum noise current is [3]:

$$\langle i_Q^2 \rangle = 2\,qI_p BM^2 F(M) \tag{8.12}$$

where

I_p = average value of the photocurrent,
q = electronic charge,
M = average of the statistically varying avalanche gain,
$F(M)$ = noise figure associated with the random nature of the avalanche process. A reasonable approximation to $F(M)$ is M^x, where $0 \leq x \leq 1.0$ and depends on the material; for silicon, $x = 0.5$, and GaInAs, $x = 0.75$.

The photodiode dark current is that which continues to flow through the bias circuit of the device when no light is incident on the photodiode. This current is a combination of bulk and surface currents. The bulk dark current arises from electrons or holes that are thermally generated in the *p-n* junction. These carriers are accelerated by the high electric field present at the *p-n* junction, and therefore multiplied by the avalanche gain mechanism. The mean square value of this current is [3]:

$$\langle i_{DB}^2 \rangle = 2qI_D BM^2 F(M) \tag{8.13}$$

where I_D is the primary unmultiplied detector bulk dark current. The surface dark current is also referred to as a surface leakage current, or simply leakage current. It is dependent on surface defects and area. The surface dark current is not affected by the avalanche gain, which is a bulk effect. The mean square value of the surface dark current is [3]:

$$\langle i_{DS}^2 \rangle = 2\,qI_L B \tag{8.14}$$

where I_L is the surface leakage current.

Let us assume that a sinusoidally modulated signal of optical power:

$$P(t) = P_0(1 + m\cos\omega t)^2 \tag{8.15}$$

is incident on the photodetector. Here, m is the optical modulation index or depth, ω is the frequency and P_0 is the average optical power. As a result a primary photocurrent, $i_{ph}(t)$, is generated, which consists of a dc value I_p and a signal component $i_p(t)$. The dc value, which is the average photocurrent due to the average optical power, is

$$I_p = R_0 P_0 \tag{8.16}$$

where R_0 is the responsivity of the photodetector.

The signal component gives rise to a mean square signal current:

$$\langle i_S^2 \rangle = \langle i_p^2(t) \rangle M^2$$
$$= \frac{m^2}{2} I_p^2 M^2 \tag{8.17}$$

The power signal to noise ratio S/N at the input of the amplifier is defined as

$$\frac{S}{N} = \frac{\text{Signal Power from Photocurrent}}{\text{Photodetector Noise Power} + \text{Thermal Noise Power due to the Load Resistor}} \tag{8.18}$$

Substituting (8.11) through (8.17) in (8.18) yields

$$\frac{S}{N} = \frac{(mR_0P_0M)^2/2}{2q(I_p + I_D)BM^2F(M) + 2qI_LB + 4KTB/R_1} \tag{8.19}$$

As a numerical example, consider a crosspoint optoelectronic switch with the following parameters:

$m = 0.8$
$R_0 = 0.7$ A/W
$P_0 = 100$ μW
$M = 20$
$I_p = 0.070$ mA
$I_D = 0.3$ nA
$I_L = 0.2$ nA
$B = 500$ MHz
$R_1 = 50$ Ω
$T = 300$ K
$F(M) = M^x$, where $x = 0.5$ for silicon APD.

The computed S/N using (8.19) is 45.0 dB.

There exists an optimum value of M which maximizes the signal-to-noise ratio. The optimum M is obtained by differentiating (8.19) with respect to M, equating the result to zero, and solving for M. This yields

$$M_{opt}^{x+2} = \frac{2qI_L + 4kT/R_1}{xq(I_p + I_D)} \qquad (8.20)$$

for $m = 1$ and $F(M) = M^x$.

The sensitivity of the optoelectronic crosspoint switch can be described in terms of the minimum detectable optical power P_{min}. This is the optical power necessary to produce a signal-to-noise ratio of one when the avalanche gain is set equal to its optimum value, as described above. P_{min} is expressed as [11]:

$$P_{min} = \frac{2h\nu}{M_{opt}e\eta} \frac{kTB^{1/2}}{R_1} \qquad (8.21)$$

Considering the previous numerical example, (8.20) yields M_{opt} of 5, and the corresponding P_{min} from (8.21) is 1.76×10^{-7} W. In executing, the analysis, the assumption made is that the postdetection amplifier is ideal, and hence contributed zero noise power. However, in a practical circuit, the amplifier noise will degrade the minimum detectable signal power level. In general, the amplifier noise output power can be expressed as

$$P_{no} = GkT_AB \qquad (8.22)$$

where

G = power gain,
T_A = effective input noise temperature.

The noise power as given by (8.22) can be referred to the input of the amplifier by dividing by G, thus yielding

$$P_{ni} = kT_AB \qquad (8.23)$$

Thus, the total noise power, P_{nt} at the amplifier input is the sum of P_{ni} and the thermal noise due to the photodetector load resistance R_1; that is,

$$P_{nt} = kTB + KT_AB$$

$$P_{nt} = k(T + T_A)B$$
$$= kT_eB \qquad (8.24)$$

where T_e is the equivalent noise temperature at the input of the amplifier following the photodiode.

Further, the amplifier noise temperature T_A is related to its noise figure F by the well known relation:

$$F = 1 + T_A/290 \qquad (8.25)$$

As a numerical example, if the postdetection amplifier discussed earlier has a noise figure of 3.0 dB, then the amplifier noise temperature T_A as determined from (8.25) is 299. This, together with T of 300, yields an equivalent noise temperature T_e of 599. Substituting the above value of T_e in (8.21) yields P_{min} of 2.5×10^{-7} W, which is higher than what was determined earlier for the ideal case.

The signal-to-noise ratio of the experimental optoelectronic crosspoint switch discussed in Section 8.4 varies between 40 to 50 dB for an incident optical power of 10 μW [1]. This is comparable to the measured signal-to-noise ratio in the case of the 20 $\times$ 20 microwave switch matrix, which is about 50 dB for an input signal level of -10 to -20 dBm [10].

8.5.3 Radiation Harness

The optoelectronic switch matrix is expected to function on a geosynchronous satellite. In this orbit, the optoelectronic circuits, if unshielded, would be typically subjected to natural ionizing radiation at the rate of about 10^5 to 10^6 RADS(Si)/year [12]. RAD (radiation absorbed dose) is a unit which corresponds to 100 ergs of energy deposited per gram of the material. Over the seven-year life expectancy of a typical geosynchronous satellite, the accumulated dose can be moderately high. Consequently, the effect of ionizing radiation on the propagation characteristics of optical fibers and on the performance of laser diodes and photodiodes is important.

In general, ionizing radiation increases the attenuation of the optical fibers, and this increase is a function of the accumulated ionizing radiation dose and temperature. Table 8.2 presents the additional attenuation at room temperature of several commercial optical fibers in a steady-state environment, as well as in a transient pulse environment, both of which subject the fiber to an accumulated ionizing radiation dose of 1 MRAD(Si). The table also presents the recovery of these fibers after being removed from the radiation field. In general, fiber without dopants in the core offer greater resistance to radiation damage. An interesting phenomenon observed in these experiments is that moderate levels of optical power propagating through the fiber during exposure to radiation helps in reducing the attenuation. This effect is known as *photobleaching* [13].

Table 8.2
Effect of Ionizing Radiation on the Attentuation of Optical Fibers [12]

Fiber Type and Part Number	Core Material	Core Diameter (μm)	Steady-State Radiation of 1 MRAD		Pulsed Radiation	
			Max Attenuation (dB/m)	Recovery 1 hour after Exposure (percent)	Attenuation 50 mS after 1 MRAD Pulse (dB/m)	Attenuation 50 mS after 14 MRAD Pulse (dB/m)
Dianichi Nippon SM200UV	Pure Silica, Low OH	200	0.25	60	0.2	0.7
Raychem WF00226	Pure Silica, High OH	200	0.2	17	—	0.8
Corning 1508	Ge-doped Silica	100	1.1	29	9.9	—
Spectran SG840-200R	Ge-doped Silica	200	0.9	39	0.2–0.4	0.3–0.5

Laboratory accelerated life test of the radiation response of laser diode structures realized in ternary and quarternary III–V compound semiconductors, show both gamma-ray irradiation as well as neutron fluence induces damage. However, a major portion of the gamma-ray irradiation induced damage can be removed by annealing, which is achieved by forward biasing the laser diode [14]. However, when exposed to neutron fluence, the recovery is rather small [14], and hence is considered to be a more serious type of damage. Neutron (n) fluence in excess of 10^{14} n/cm^2 is adequate to damage the lattice of the semiconductor crystal. The primary effect of this damage is to reduce the minority carrier lifetime, which in turn provides an alternative nonradiative recombination path for the carriers. The result of nonradiative recombination is generation of heat instead of light.

From a circuit point of view, we can observe that the neutron fluence increases the threshold current of the laser diode, but has insignificant effect when lasing is well underway. This is because the minority carrier lifetime in GaAs, when operated near threshold, is on the order of 1 to 10 ns. However, under intense stimulated emission, the lifetime is on the order of 1 to 10 ps [15]. Therefore, a much larger concentration of defects is required to influence the radiative recombination rate through competing non-radiative recombination. In view of the above, care should be taken to select a laser diode with low threshold current and high maximum allowable operating current.

In the case of photodiodes, in general, the reduction in the minority carrier lifetime reduces the diffusion length [14]. The reduction in the diffusion length causes the optically generated minority carriers created farthest from the junction to recombine before diffusing into the depletion region. In addition, there is also a greater loss of optically generated minority carriers due to recombination while traveling through the depletion region. The loss of minority carriers lowers the responsivity of the photodiode. In addition the leakage current is also increased by generation-recombination sites, due to radiation damage, within the depletion region of the photodiode. The increase in the leakage current creates an increase in the noise current, which in turn degrades the signal-to-noise ratio [3]. Finally, there is also some indication that the $1/f$ noise of an APD also increases with radiation [16]. The radiation hardness of the photodiodes can be improved by constructing the devices out of III–V compound semiconductor material. At 0.82 μm wavelength, these materials have relatively large optical absorption coefficients. The large optical absorption coefficient allows for the fabrication of a very thin active region without affecting responsivity. Because ionizing radiation uniformly generates electron-hole pairs throughout the active volume, the smaller is the active volume, the smaller will be the radiation generated current. Also, the III–V compound semiconductor material also allows the fabrication of heterojunctions which have built-in electric fields. These internal fields prevent the radiation generated carriers in the substrate region from diffusing into the active region. Based on the above rationale, double-heterostructure AlGaAs-GaAs p-n junction photodiodes have been fabricated and found to have negligible leakage current when exposed to ionizing radiation [13]. In general, this area has remained largely unexplored, particularly the effect of radiation damage on the RF performance of the above devices.

GaAs MMICs have made tremendous strides in the past few years [17]. Hence, future microwave switch matrix to a large extent will incorporate MMICs to achieve various circuit functions. Therefore, appropriate here for comparison is to discuss briefly the effect of radiation on MMICs. The electron irradiation hardness level for GaAs MESFETs under normal bias is about 5×10^7 RAD(Si) [18]. This is a high tolerance value and acceptable for most applications. The neutron irradiation hardness level for drain current degradation is 3×10^{14} n/cm^2 and for gain degradation is 3×10^{14} n/cm^2 [18]. The degradation point being 20%. The mechanism of degradation here is carrier removal. The above statistics show that the radiation hardness of optoelectronic devices and microwave solid state devices are about the same level.

8.5.4 Reliability

The two basic failure mechanisms that limit laser diode life are catastrophic and gradual degradation. The first depends on the optical flux density and pulsewidth, resulting in facet damage, and the second is mainly a function of the current density, duty cycle, and details of the laser fabrication process. Catastrophic damage can be minimized by antireflection films on the laser facet. Gradual degradation can be slowed by operating the laser diode at a lower temperature. Experiments have suggested that the room-temperature CW operating life of the best laser diodes exceeds 100,000 hours [19]. However, these results are not valid for pulsed operation. Under pulsed operation, the time to failure is about an order of magnitude lower than the best CW lifetime [20]. A more stringent and valid method of evaluation is continuous bit error rate testing of pulse code modulated transmitters. This technique reveals any short-duration instabilities that will seriously affect system performance, but which are not found by CW or pulsed testing [20]. Optoelectronic switching is a relatively new concept, and so estimates of the laser's contribution to the total switch matrix lifetime is unavailable at present.

Reliability of GaAs MESFET switched amplifiers is the factor that determines the lifetime of the microwave switch matrix. The possibility of failure of a crosspoint in the on-state is eliminated by using two series switches with separate drivers at each crosspoint. The possibility of failure of a crosspoint in the off-state is eliminated by a wrap-around method, which uses extra rows and columns of crosspoints. The extrapolated lifetime for the microwave switch matrix is greater than 10^9 hours [21].

REFERENCES

1. MacDonald, R.I., and E.H. Hara, "Switching with Photodiodes," *IEEE Quantum Electronics*, Vol. QE-16, No. 3, March 1980, pp. 289–295.
2. Hara, E.H., and R.I. MacDonald, "A Broad-Band Optoelectronic Microwave Switch," *IEEE Trans. Microwave Theory Tech.*, Vol. MTT-28, No. 6, June 1980, pp. 662–665.
3. Keiser, G., Optical Fiber Communications, New York, McGraw-Hill, 1983, pp. 127–128, 150–156, 166.
4. Kiehl, R.A., and D.M. Drury, "Performance of Optically Coupled Microwave Switching Devices," *IEEE Trans. Microwave Theory Tech.*, Vol. MTT-29, No. 10, October 1981, pp. 1004–1010.
5. Laser Diode Product Catalog, Spectra Diode Laboratories.
6. Optoelectronic Product Selection Guide, Mitsubishi Electronics America, Inc.
7. Gupta, R.K., and F.T. Assal, "Integration, Environmental, and Radiation Testing of a Microwave Switch Matrix System," *IEEE Int. Conf. Communications*, June 14–18, 1981, Denver, Vol. 1, pp. 5.4.1 to 5.4.6.
8. Lau, K.Y., and A. Yariv, "Intermodulation Distortion in a Directly Modulated Semiconductor Injection Laser," *Applied Physics Letters*, Vol. 45, No. 10, November 1984, pp. 1034–1036.
9. Darcie, T.E., R.S. Tucker, and G.J. Sullivan, "Intermodulation and Harmonic Distortion in InGaAsP Lasers," *Electronics Letters*, Vol. 21, No. 16, August 1985, pp. 665–666.

10. Saunders, A., "20 × 20 High Speed Microwave Matrix Switch," *NASA Tech. Memo.*, No. 83775, September 1984.
11. Yariv, A., "Optical Electronics," 3rd Ed., New York, Holt, Rinehart and Winston, 1985, p. 379.
12. Michal, R.J., "Radiation Effects in Optical Fiber," *Optical Technologies for Communication Satellite Applications*, SPIE, Vol. 616, 1986, pp. 292–298.
13. Wiczer, J.J., and C.E. Barnes, "Opto-electronic Data Link Designed for Applications in a Radiation Environment," *IEEE Trans. Nuclear Science*, Vol. NS-32, No. 6, December 1985, pp. 4046–4049.
14. Barnes, C.E., and J.J. Wiczer, "Radiation Effects in Optoelectronic Devices," Sandia National Laboratories, Report No. SAND-84-0771, Albuquerque, NM, May 1984.
15. Basov, N., Yu. Drozhbin, Yu. Zkharov, V. Nikitin, A. Semenov, B. Stepanov, A. Toimachev, and V. Yakovlev, "Effect of Injection Current on the Time Dependence of the Emission from GaAs Lasers," *Sov. Phys.—Solid State*, Vol. 8, 1967, 2254.
16. Swanson, E.A., E.R. Arnau, and F.G. Walther, "Measurements of Natural Radiation Effects in a Low Noise Avalanche Photodiode," *IEEE Trans. Nuclear Science*, Vol. NS-34, No. 6, December 1987, pp. 1658–1661.
17. Pucel, R.A. (ed.), *Monolithic Microwave Integrated Circuits*, New York, IEEE Press, 1985.
18. Anderson, W.T., M. Simons, A. Christou, and J. Beall, "GaAs MMIC Technology Radiation Effects," *IEEE Trans. Nuclear Science*, Vol. NS-32, No. 6, December 1985, pp. 4040–4045.
19. Willardson, R.K., and A.C. Beer (eds.), *Semiconductors and Semimetals, Vol. 14*, New York, Academic Press, 1979, pp. 151–161.
20. Tsang, W.T. (ed.), *Semiconductors and Semimetals, Vol. 22*, New York, Academic Press, 1985, pp. 184–189.
21. Prather, W.H., B.J. Cory, R.F. Wade, W.J. Taft, and R.E. Buzinski, "Wideband, High Speed Switch Matrix Development for SS-TDMA Applications," *IEEE Int. Conf. Communications*, June 14–18, 1981, Denver, Vol. 1, pp. 5.3.1 to 5.3.5.
22. MacDonald, R.I., and E.H. Hara, "The Optoelectronic Switch Matrix for On-board SS/TDMA Applications," *IEEE Int. Conf. Communications*, June 14–18, 1981, Denver, Vol. 1, pp. 15.1.1 to 15.1.5.
23. Kitazume, S., Y. Takimoto, N. Komiyama, and K. Betaharon, "Switch Matrix Development for INTELSAT SS/TDMA System," *IEEE Inter. Conf. Communications*, May 14–17, 1984, Amsterdam, the Netherlands, Vol. 2, pp. 796–799.

Chapter 9
Optoelectronic Switching and Modulation of Oscillators

9.1 INTRODUCTION

In Chapter 6, we discussed an optoelectronic microwave switch constructed from a microstrip on a semiconductor substrate. In this transmission line, an opening etched in the top strip conductor exposes the semiconductor substrate to optical illumination from a laser. The optical illumination enhances the conductivity of the semiconductor below the opening, resulting in a continuous path for the signal propagating along the line. Thus, by controlling the conductivity, switching and gating of microwave signals is possible.

In Chapter 8, we discussed an optoelectronic microwave switch constructed from an avalanche photodiode. In this case, the sensitivity of the APD to optical illumination increases several orders of magnitude as the bias voltage across the diode is switched from zero to a large negative value. This is because of the avalanche multiplication of the carriers that occurs within the device. Thus, by controlling the sensitivity, detecting, amplifying, and routing microwave signals is possible.

In future active phased arrays, each element will have its own microwave power source and will require control signals for synchronization. The use of conventional coaxial cables for control signal distribution would render the array prohibitively large, heavy, and complex. One solution to this problem is to use single-mode optical fiber and semiconductor laser diode technology. Current technology shows that semiconductor laser diodes are capable of being directly modulated at microwave frequencies, and therefore has the potential to control microwave devices. Hence, there is a need to explore the possibility for direct optical control of IMPATT and MESFET oscillators.

In this chapter, optical illumination is used to switch [1–3] as well as to modulate [14–6] a microwave oscillator. The oscillator discussed consists of a MESFET or an IMPATT diode. In the case of an IMPATT, when the diode is illuminated by a laser, the optically generated carriers enhance the reverse saturation current. The increase in the reverse saturation current alters the diode admittance. The change in the admittance is responsible for the change in the quality factor, Q, of the cavity. A change in the Q leads to a change in both the output power and frequency of oscillation. In the case of a MESFET, optical illumination causes photocapacitive, photovoltaid, and photoconductive effects in the active region of the device [3]. The photovoltaic and photoconductive effects are responsible for the change in the oscillator power while the photocapacitive effect changes the frequency of oscillation.

9.2 PHYSICAL MECHANISM OF IMPATT DIODE OPERATION

9.2.1 Unilluminated IMPATT Diode

The physical mechanism of IMPATT diode operation can be understood by considering a p^+-n-n^+ device structure, which is shown in Figure 9.1(a). Across this structure, a large dc voltage is applied such that the device is reverse biased. The resulting electric field distribution is shown in Figure 9.1(b). This electric field is less than the critical field needed for avalanche breakdown. The bias current under this condition is zero. However, an exceedingly small amount of leakage current or reverse saturation current flows.

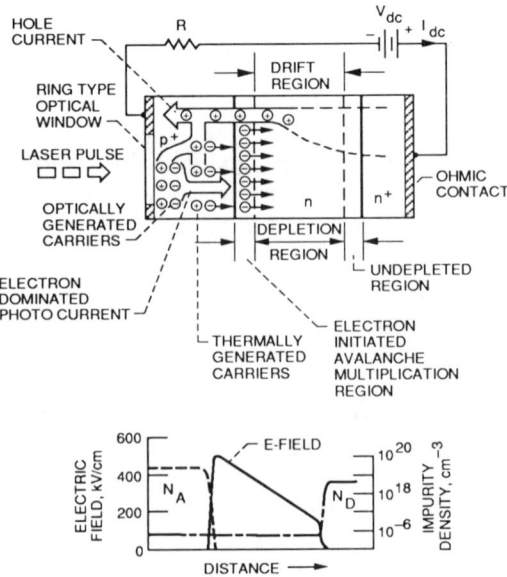

Figure 9.1 (a) A silicon p^+-n-n^+ IMPATT diode schematic; (b) doping profile and dc electric field *versus* distance.

The IMPATT diode is normally mounted in a microwave cavity for it to generate RF oscillations. The impedance of the cavity is inductive and matched to the capacitive impedance of the diode so as to form a resonant circuit. In such a circuit a noise spike will cause ringing, which for a few cycles is almost like the application of a sinusoidal ac voltage across the diode terminal (Figure 9.2(a)). This ac voltage V_{rf} is superimposed on the dc bias voltage.

If the dc bias voltage is such that it sets up a critical electric field, then, during the positive half cycle, the net electric field across the diode is greater than the critical field, and avalanche multiplication hence is initiated. The avalanche multiplication or the impact ionization rate per carrier follows the field change in phase. However, the carrier density does not follow the field change in phase because the carrier generation also depends on the number of carriers already present. Thus, at $\omega t = 0$ the carriers that are present are those that are thermally generated and correspond to the reverse saturation current. Because the reverse saturation current is small, the carrier generation due to avalanche is also small. At $\omega t = \pi/2$, the ac voltage reaches its maximum. However, the carrier density has not yet reached its maximum, insofar as the carrier generation rate is still increasing. The maximum carrier density is

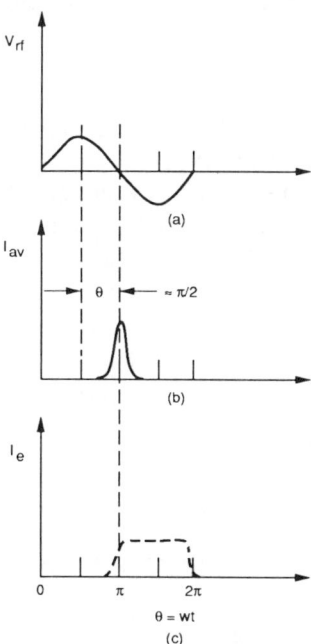

Figure 9.2 Phase relation without optical illumination: (a) RF voltage *versus* time; (b) avalanche current *versus* time; (c) external current *versus* time.

reached approximately at $\omega t = \pi$. When the ac voltage changes sign and subtracts from the dc voltage, the field falls below the critical field and the generation rate starts decreasing, causing the current to decrease. Thus, the ac variation of the injected carrier density or the avalanche current I_{av} is a short pulse, and lags the peak ac voltage by about $\pi/2$ as shown in Figure 9.2(b). This time lag is known as the *avalanche delay,* denoted as θ. The injected carriers then enter the drift region of length L, where they travel at saturation velocity V_s, introducing the transit time delay T. The quantities L, V_s, and T are related as follows:

$$L = V_s T \qquad (9.1)$$

The frequency, f, of the ac voltage is

$$f = 1/2T \qquad (9.2)$$

The induced external current waveform is shown in Figure 9.2(c). The external current waveform has a fundamental Fourier component that is 180° out of phase with the voltage waveform. This constitutes an ac negative resistance.

The efficiency of the IMPATT diode is expressed as

$$\eta = P_{ac}/P_{dc} \qquad (9.3)$$

A more detailed discussion of IMPATT diode operation, admittance, efficiency, and power output is available elsewhere [19].

As a numerical example, a silicon single-drift IMPATT didoe has a drift region length of 4.5 μm, drift velocity of 10^7 cm/s, maximum operation volatage of 100 V, maximum operating current of 230 mA, and efficiency of 11%. The maximum CW power output from the idode is

$$\begin{aligned} P_{ac} &= \eta P_{dc} \\ &= 0.11 \cdot 100 \cdot 0.23 \\ &= 2.53 \text{ W} \end{aligned}$$

The transit time frequency of the IMPATT diode is

$$\begin{aligned} f &= 10^7/2 \cdot 4.5 \cdot 10^{-4} \\ &= 11.11 \text{ GHz} \end{aligned}$$

9.2.2 Illuminated IMPATT Diode

The experimental technique for illuminating the IMPATT diode is subsequently discussed in Section 9.5. However, for the present, let us say that the diode is optically illuminated by a laser. The optical illumination generates carriers that increase the reverse saturation current. The increase in the reverse saturation current causes the avalanche current I_{av} to attain a peak value in a much shorter time, thereby causing the avalanche delay θ to reduce. That is, the phase difference between the peak V_{rf} and I_{av} lags by an angle less than $\pi/2$. This is illustrated in Figure 9.3(a) and (b). This reduction in phase introduces a resistive as well as a change in the reactive component in I_{av}. The resistive component causes the oscillations to cease, or a reduction in the output power, while the change in the reactive component causes a shift in the frequency of oscillation [2]. The external current waveform I_e (Figure 9.3(c)) also lags the ac voltage by an angle of less than π.

9.3 IMPATT DIODE OSCILLATOR DESIGN

9.3.1 Reduced Height Waveguide Cavity

Figure 9.4(a) illustrates an IMPATT diode oscillator mount that uses a reduced height rectangular waveguide. In this mount, the IMPATT diode is located in a circular

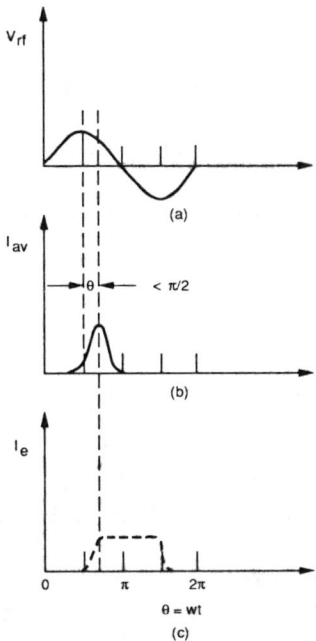

Figure 9.3 Phase relation with optical illumination: (a) RF voltage *versus* time; (b) avalanche current *versus* time; (c) external current *versus* time.

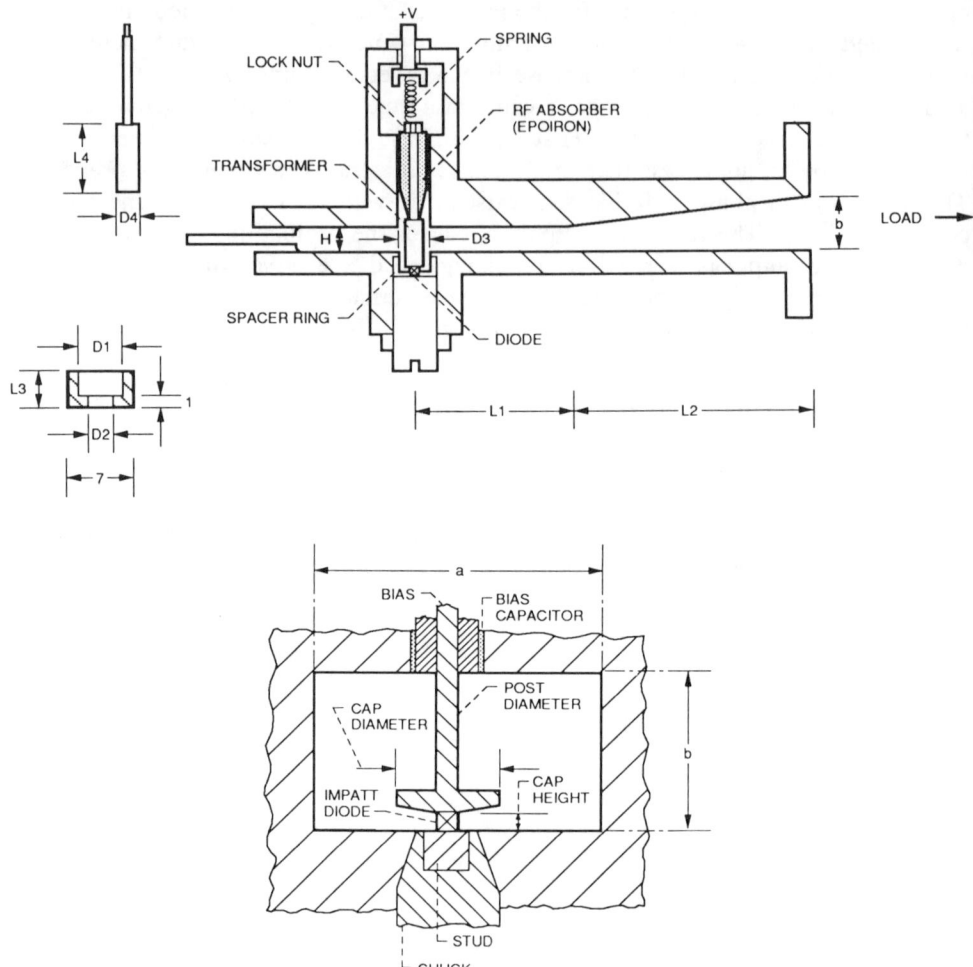

Figure 9.4 Schematic of IMPATT oscillators: (a) reduced height waveguide cavity; (b) standard height waveguide cavity with resonant cap.

recess in the bottom broad wall of the waveguide. The bias to the diode is applied via a 50 Ω coaxial line, which passes through the waveguide at the center of the top wall, and makes contact to the diode. Impedance matching between the 50 Ω coaxial line and the diode is accomplished by a quarter-wave stepped-impedance transformer. Further, the coaxial line is impedance-matched to a standard waveguide or load by a tapered height waveguide section. The RF absorbing material in the bias circuit helps prevent spurious microwave oscillations.

The frequency of oscillation is determined by the geometry of the transformer, and precise frequency tuning is done by a movable short circuit. The disadvantage of this mount is that it is very expensive, as several precision parts are required. Furthermore, interchangeability of diodes is more difficult. However, this type of mount is capable of providing a very broad range of tuning and impedance by change of transformer. Table 9.1 presents the mount dimensions [7] for several waveguide frequency bands.

Table 9.1
IMPATT Oscillator Dimensions (mm)

-Frequency (GHz)	Waveguide	L_1	L_2	L_3	L_4	D_1	D_2	D_3	D_4	H
6–6.5	WR-159	$n\lambda_g/2$	$n\lambda_g$	15–5	25–18	6.5	4.0	6.5	6.0–5.0	6.0
7–8.2	WR-137	$n\lambda_g/2$	$n\lambda_g$	13–4	20–10	6.5	3.8	6.5	6.0–5.0	5.0
8–10	WR-112	$n\lambda_g/2$	$n\lambda_g$	10–3	16–9	6.4	3.6	6.4	6.0–5.0	4.0
10–12.4	WR-90	$n\lambda_g/2$	$n\lambda_g$	8–2	12–8	6.0	3.4	6.0	4.6–5.0	3.0
12–14	WR-75	$n\lambda_g/2$	$n\lambda_g$	7–1.5	10–5	5.5	3.4	5.5	4.4–4.0	2.0
13–18	WR-62	$n\lambda_g/2$	$n\lambda_g$	6–1.5	8–3.6	5.0	3.0	5.0	4.0–3.6	1.5
17–22	WR-51	$n\lambda_g/2$	$n\lambda_g$	6–1.0	6–3.6	4.0	2.4	4.0	3.4–2.8	1.0
22–33	WR-34	$n\lambda_g/2$	$n\lambda_g$	4–0.5	6–2.5	3.1	2.4	3.1	2.5–2.0	1.0
27–40	WR-28	$n\lambda_g/2$	$n\lambda_g$	2–0.5	5–2.5	3.1	1.2	3.1	2.5–2.0	1.0

Source: Microwave IMPATT Diode Data Sheet (ND 8 Si CW IMPATT), California Eastern Laboratories (NEC Corporation).

9.3.2 Standard Height Waveguide Cavity with Resonant Cap

Figure 9.4(b) illustrates a standard waveguide with a resonant cap structure for IMPATT diode oscillators. The resonant cap structure is also called a "top hat." The top hat structure is a method of tuning and coupling an IMPATT diode in a waveguide using a circular disk, which is mounted on top of the diode package. The zone between the disk and the ground plane acts as a radial transmission line. Experiments have shown that the cap radius is roughly a quarter-wavelength [8]. In this mount, the operating frequency can be tuned over 20% bandwidth by varying the cap height by 10% of the waveguide height, the optimum height being 20 to 25% of the waveguide height. The post diameter and cap thickness are roughly 5 to 10% of the waveguide height [9].

9.4 EXPERIMENTAL DEMONSTRATION OF ENHANCEMENT AND QUENCHING OF IMPATT DIODE OSCILLATOR UNDER OPTICAL ILLUMINATION

In the experiment described here [1], a GaAs IMPATT diode is mounted in a tunable waveguide cavity. The light from a GaAlAs laser diode is coupled to the active region of the diode by an optical fiber.

To demonstrate quenching, the IMPATT diode is biased above threshold and the cavity is tuned to optimize the microwave output power. Illumination by an optical pulse causes the IMPATT diode to shut off for a time interval corresponding to the light pulse duration. Figure 9.5(a) presents an oscillogram showing the quenching of the microwave oscillations.

To demonstrate enhancement, the cavity is detuned slightly, so that the cavity favors some other frequency of oscillation which has a higher threshold. The IMPATT diode, when illuminated by an optical pulse, is triggered into oscillations. Figure 9.5(b) presents an oscillogram showing the enhancement of the microwave oscillations.

Laboratory experiments [10] have also shown that the rise and fall time of the RF pulse is about the same as that of the optical pulse, which is on the order of a fraction of a nanosecond. Further, RF pulses which are a few nanoseconds wide can

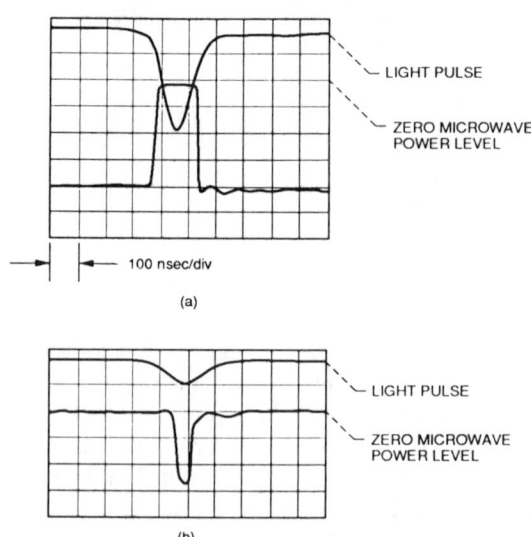

Figure 9.5 Oscillogram showing the (a) quenching and (b) enhancement.
Source: Yen, H.W., M.K. Barnoski, R.G. Hunsperger, and R.T. Melville, "Switching of GaAs IMPATT Diode Oscillator by Optical Illumination," *Applied Physics Letters*, Vol. 31, No. 2, July 1977, pp. 120–122. Reprinted with permission.

be obtained by this technique. Finally, a few watts of optical power is sufficient either to enhance or quench the IMPATT oscillations. This magnitude of optical power can be easily obtained from a semiconductor laser diode.

However, an electronically controlled, pulsed IMPATT oscillator uses special modulator circuits to provide a bias current pulse. Typical rise and fall time of the RF pulse that can be obtained with this technique is on the order of a few nanoseconds [11]. However, when compared to the optical technique, it is an order of magnitude slower. The corresponding RF pulse width can be varied typically in the range of a few hundred nanoseconds [11]. Again, the optical technique is superior, as it can provide a pulsewidth which is two orders of magnitude smaller.

9.4.1 Frequency Chirp

In radar applications, high peak power and narrow pulsewidths are essential for range and angular resolution, respectively. IMPATT diodes with pulsed bias current are capable of meeting these requirements. However, during the pulsing operation, the IMPATT diode junction temperature increases because of the heat generated due to the circuit losses. For a uniform current pulse, the diode junction is at a low temperature at the beginning of the pulse and gradually heats up, depending on its thermal time constant. This temperature variation depends on current density, junction area, and thermal resistance. As the diode warms up during the pulse, the device impedance changes. Consequently, the frequency of oscillation also changes. This change in the frequency is termed *chirp*.

Because the diode impedance is dependent on the bias current, the frequency variation caused by thermal effects can be compensated by changing the operating current density. This technique [11] is illustrated in Figure 9.6. For a uniform current pulse, the oscillator frequency has a downward chirp. By providing an upward ramp on the current pulse, the amount of chirp can be decreased. A continuous increase in the ramp slope will reach a point where the thermal and current effects cancel each other and little chirp is present. Further increase in the ramp slope beyond this point will cause the frequency to chirp upward. Thus, by controlling the current waveform, the frequency chirp characteristics can be controlled.

Laboratory experiments [2] on quenching of IMPATT oscillations with optical illumination have shown that quenching a small portion of an RF pulse is also possible, either the leading edge or the trailing edge. An interesting observation made during such an experiment is that if the trailing edge of the RF pulse is inhibited, the frequency chirp of the IMPATT diode RF pulse becomes reduced.

9.5 IMPATT DIODE STRUCTURES WITH ETCHED OPTICAL WINDOW

As described in Section 9.4, the IMPATT diodes are either edge-illuminated or contact-illuminated. This technique has the disadvantage of poor coupling efficiency

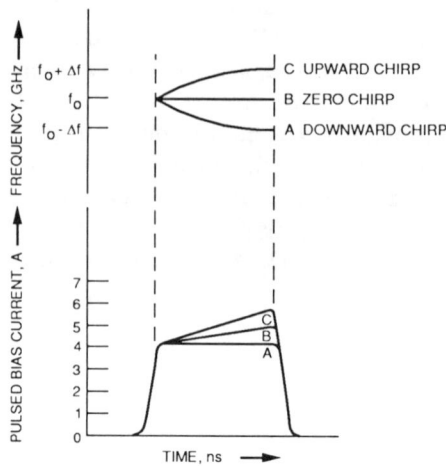

Figure 9.6 Frequency chirp characteristics of a pulsed IMPATT oscillator (Δf is the chirp frequency).

between the optical source and the active layer of the IMPATT diode. One technique of improving the coupling efficiency is to provide an etched optical window at the top contact of the diode. Figure 9.7 schematically illustrates two possible IMPATT diode structures with an optical window or a ring contact. In these devices, the optical window may be offset from the device diameter to provide sufficient area for bonding, while maximizing the window area.

The quality factor, Q, the contact spreading resistance, the frequency of oscillation, and the CW power output of the IMPATT diodes, with ring contact as well as conventional dot contact, were measured and compared [12]. The measurements indicated that there is no difference between the results of the two cases. Hence, an

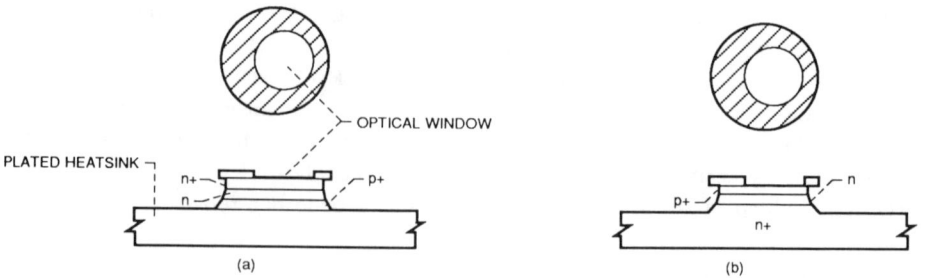

Figure 9.7 Silicon IMPATT diode structures for photoexcitation: (a) flip chip diode; (b) top-mounted diode (after [13], p. 212).

etched optical window could be used for illuminating an IMPATT diode without regrading its microwave performance.

9.6 EFFECT OF HOLE *VERSUS* ELECTRON PHOTOCURRENT ON SILICON IMPATT DIODE OSCILLATOR POWER AND FREQUENCY

Section 9.4 describes enhancement in the IMPATT diode oscillator signal amplitude when optically illuminated. This change is attributed to the change in the negative admittance, which is caused by a change in the magnitude of the reverse saturation current. Because the ionization coefficients for electrons and holes are different in silicon, the composition of the reverse saturation current in terms of electron and hole currents will decide the degree of control.

The structure shown in Figure 9.7(a) has its junction side down (i.e., plated heat sink or flip chip). This structure when illuminated from the n^+ side generates hole-dominated photocurrent. Conversely, the structure shown in Figure 9.7(b) has its junction side up (i.e., top-mounted). This structure when illuminated from the p^+ side generates electron dominated photocurrent. Thus, electron multiplication dominates in the top-mounted device, and hole multiplication dominates in the flip chip device under optical illumination.

IMPATT diode oscillators fabricated with these devices show that the power output and frequency of oscillation are an order of magnitude more sensitive when the photocurrent is electron-dominated than when it is hole-dominated [13]. In other words, for the same photocurrent magnitude, larger amplitude modulation depths of the RF power of an IMPATT oscillator are achieved when the photocurrent is electron-dominated than when the photocurrent is hole-dominated.

9.7 PHOTOCURRENT EFFECTS ON SILICON IMPATT OSCILLATOR NOISE

Experiments have shown that an increase in the reverse saturation current leads to a reduction in the oscillator noise [14, 15]. One method of increasing the reverse saturation current is by optically illuminating the IMPATT diode. Noise measurements on IMPATT diode oscillators show that the AM noise increases with the photocurrent, whether the devices are fabricated as flip chip or top-mounted. However, the increase is greater for top-mounted IMPATT diodes (Figure 9.7(b)) with electron-dominated photocurrent.

The FM noise at small bias currents is observed to increase as the photocurrent increases in the case of top-mounted IMPATT diodes. However, the FM noise at small bias currents is independent of the photocurrent in the case of flip chip IMPATT diode. Thus, the noise increase is greater with electron-initiated photocurrent in silicon devices, due to the higher ionization coefficient of electrons [15].

9.8 FREQUENCY MODULATION OF A SILICON IMPATT DIODE OSCILLATOR BY OPTICALLY GENERATED CARRIERS

To frequency-modulate an IMPATT diode oscillator, the modulating signal at a frequency f_m is first made to intensity-modulate a semiconductor laser diode. The emitted optical signal from the laser then illuminates the IMPATT diode. As a result, the avalanche delay θ varies, depending on the intensity of the optical illumination. This variation in the avalanche delay introduces a varying reactive component in I_{av}, as explained in Section 9.2.2. The reactive component is responsible for the change in the oscillator frequency. Experiments with a silicon IMPATT diode show that the highest modulation frequency, f_m, achieved is a few MHz [5]. Beyond this limit, the frequency deviation of the oscillator decreases drastically. This limitation is due to the low diffusion coefficient of the minority carriers generated in the n^+ region when a portion of the optical power is absorbed in the substrate.

9.9 GaAs MESFET OSCILLATOR DESIGN

Microwave oscillators using a GaAs MESFET normally operate under large-signal conditions. However, the only parameter that changes from small-signal to large-signal operation is the device transconductance. The rest of the parameters affecting the oscillation frequency scarcely change. Therefore, a large-signal oscillator with a GaAs MESFET can be designed by using small-signal scattering parameters (S-parameters) [16].

Microwave oscillators in general can be constructed either with series feedback or parallel feedback elements. In this section, an example of an oscillator with series feedback will be discussed.

9.9.1 Series Feedback Oscillator

Figure 9.8 shows the lumped-element circuit of the oscillator. The reactance between the source and the ground terminals provides the feedback. The element values for this oscillator can be obtained as explained in the following example.

Let us suppose that an oscillator is required at 3 GHz with maximum efficient output power gain of 5 dB. The maximum efficient power gain, G_{osc}, of the oscillator is given by the expression [17]:

$$G_{osc} = (G_0 - 1)/\ln(G_0) \qquad (9.4)$$

where G_0 is the small-signal gain of the device as an amplifier and is equal to $|S_{21}|^2$. The stability factor K is given by the expression [17]:

$$K = \frac{1 + |\Delta|^2 - |S_{11}|^2 - |S_{22}|^2}{2|S_{12}| |S_{21}|} \qquad (9.5)$$

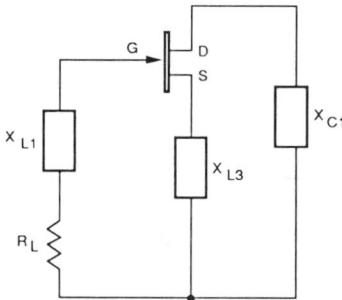

Figure 9.8 Series feedback GaAs MESFET oscillator circuit.

where

$$\Delta = S_{11} S_{22} - S_{21} S_{12} \tag{9.6}$$

Further, let us suppose that the device selected has the following small-signal, common-source S-parameters (magnitude and angle) at 3 GHz:

$$S_{11} = 0.589, -153°$$

$$S_{21} = 3.248, 20°$$

$$S_{12} = 0.061, 15°$$

$$S_{22} = 0.634, -100°$$

The first step is to compute K, G_0, and G_{osc}, which are 0.13, 10.55, and 6.1 dB, respectively. These computations show that the device meets the condition for oscillation, which is K should be less than unity. Further, the power requirement of 5 dB is also met.

The oscillator circuit element values are given by the following equations [17,18]:

$$R_L = D_1 + F_r D_3 + F_i D_4 \tag{9.7}$$

$$X_{L1} = D_2 - (1 + F_r)(D_4 + D_3 F_r / F_i) \tag{9.8}$$

$$X_{C2} = -\frac{D_3(1 + F_r)}{F_i} - D_4 \tag{9.9}$$

$$X_{L3} = \frac{D_3 F_r}{F_i} + D_4 \tag{9.10}$$

where

$$D_1 = -\text{Re}(Z_{11} + FZ_{12}) \tag{9.11}$$

$$D_2 = -\text{Im}(Z_{11} + FZ_{12}) \tag{9.12}$$

$$D_3 = -\text{Re}(Z_{21} + FZ_{22}) \tag{9.13}$$

$$D_4 = -\text{Im}(Z_{21} + FZ_{22}) \tag{9.14}$$

$$F = F_r + jF_i$$

$$= \frac{Z_{21} - AZ_{11}}{AZ_{12} - Z_{22}} \tag{9.15}$$

$$A = A_r + jA_i$$

$$= -\frac{(Y_{21} + Y_{12}^*)}{2\,\text{Re}\,Y_{22}} \tag{9.16}$$

The second step is to convert the S-parameters into equivalent impedance (Z) and admittance (Y) parameters. The resulting parameters are:

$$Z_{11} = 0.412044410856 - j0.259837431991$$

$$Z_{12} = 0.062431980803 - j0.028823928719$$

$$Z_{21} = 3.445360150550 - j1.239188637140$$

$$Z_{22} = 0.513821796202 - j0.868217094011$$

$$Y_{11} = 1.275198243840 + j2.416765797640$$

$$Y_{12} = 0.021995262074 - j0.184948538221$$

$$Y_{21} = 2.024989062520 - j9.708204899660$$

$$Y_{22} = 0.015464175651 - j1.319322903310$$

The third step is to compute the A and F factors, which are

$$A = -66.1847217315 + j307.913482623$$

$$F = -6.78511802834 + 0.936108757014$$

The fourth step is to compute the D factors, which are

$$D_1 = -0.015418384447$$

$$D_2 = 0.005820549669$$

$$D_3 = -0.771764242483$$

$$D_4 = -5.132759902920$$

The last step is to compute the element values, which are

$$R_L = 0.42 \ \Omega$$

$$X_{L1} = 2.67364892095$$

$$L_1 = 0.14 \text{ nH}$$

$$X_{C2} = 0.36328498949$$

$$C_2 = 146 \text{ pF}$$

$$X_{L3} = 0.46115366328$$

$$L_3 = 0.03 \text{ nH}$$

In general, the MESFET oscillator can be operated either in the optically-on mode or in the optically-off mode. In the optically-on mode the MESFET is biased close to cut-off and enhancement of the oscillations occurs with light. While in the optically-off mode, the MESFET is biased close to saturation, and illumination produces quenching of the oscillations. The first technique has the advantage of lower dc power dissipation because the device is biased close to cut-off. Experimentally, for maximum output power, the optimum gate bias voltage has been observed to be about half the pinch-off voltage [6].

9.10 EXPERIMENTAL DEMONSTRATION OF SWITCHING OF GaAs MESFET OSCILLATOR UNDER OPTICAL ILLUMINATION

The experimental oscillator [3] consists of a GaAs MESFET operating in the optically-on mode. The MESFET is illuminated by a pulsed optical beam, and the waveform of the drain current is monitored. The waveform shows that both the leading and the trailing edges of the transient drain current response have more than one time component as shown in Figure 9.9. The leading edge has a fast component with a rise time t_1, which is the same as that of the pulsed laser driving current, and a duration on the order of a few ns. The slow component of the leading edge has a rise time t_2, which is on the order of a μs. Similarly, the trailing edge fall time has a fast component that has a fall time t_3, which follows the laser pulse and has a duration on the order of a few ns. This is followed by a slow component, which has a fall time t_4 on the order of a few μs, and finally a long decay time t_5 on the order of a few ms.

When the GaAs MESFET is optically illuminated several light-induced phenomena occur. They are photocapacitive effects in the gate-to-source depletion region, photovoltaic effects at channel-to-substrate interface layer and photoconductive effects in the channel [3]. The photocapacitive and photovoltaic effects are due to the capture of holes by hole traps, which arise from the Cr doping, defects in the material and other impurities. The fast response times, t_1 and t_3, are attributed to

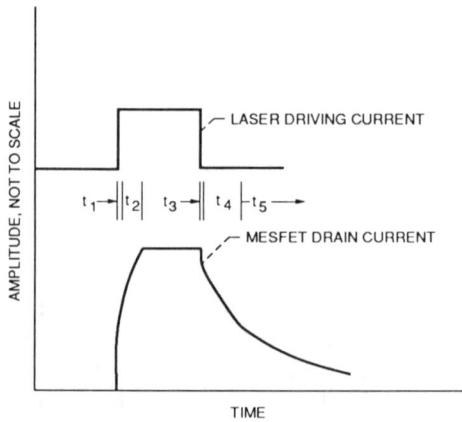

Figure 9.9 Typical optically induced transient drain current response shape of GaAs MESFET oscillator.

photoconductive effects in the channel. The two long and different response times, t_2 and t_4, indicate that two kinds of hole traps with different activation energies are contributing to the optical response. Hence, several seconds must elapse after the light is turned off for the traps to emit holes and the oscillation to return to its normal state. Finally, we can observe that, as the laser PRF (pulse repetition frequency) is increased to a few MHz, the response of the MESFET oscillator decreases drastically, due to the slow emission time of the hole traps.

9.11 FREQUENCY MODULATION OF A GaAs MESFET OSCILLATOR BY OPTICALLY GENERATED CARRIERS

The technique is similar to that presented for the case of an IMPATT diode in Section 9.8. In the MESFET, as a result of the photocapacitive effects, the gate-to-source capacitance changes. The change in capacitance alters the frequency of oscillation. The maximum frequency deviation in the case of MESFETs also is limited to a few MHz [6]. This limitation is due to the slow emission time of the hole traps, as explained in Section 9.10. A comparision of the noise performance of GaAs MESFET oscillators with other solid-state sources has been presented in Section 7.7.2 of Chapter 7 and the results have been summarized in Table 7.1.

REFERENCES

1. Yen, H.W., M.K. Barnoski, R.G. Hunsperger, and R.T. Melville, "Switching of GaAs IMPATT Diode Oscillator by Optical Illumination," *Applied Physics Letters,* Vol. 31, No. 2, July 1977, pp. 120–122.
2. Gerlach, H.W.A., and R. Wellman, "The Behavior of a Pulsed Millimeter Wave (70 GHz) IMPATT Diode Oscillator During Laser Illumination," *IEEE Int. Microwave Symp. Digest,* 1980, pp. 70–72.
3. Sun, H.J., R.J. Gutmann, and J.M. Borrego, "DC and Pulse-Light Illuminated Optical Responses of Microwave GaAs-MESFET Oscillators," *IEE Proc.,* Vol. 131, Pt. I, No. 1, February 1984, pp. 31–37.
4. Kiehl, R.A., "Optically Induced AM and FM in IMPATT Diode Oscillators," *IEEE Trans. Electron Devices,* Vol. ED-27, No. 2, February 1980, pp. 426–432.
5. Chiu, C., and J. Freyer, "Frequency Modulation of IMPATT Diodes by Optical Illumination," *IEE Proc.,* Vol. 131, Pt. I, No. 1, February 1984, pp. 28–30.
6. Loriou, B., J. Guena, and J.F. Sautereau, "Optically Frequency-Modulated GaAs MESFET Oscillator," *Electronics Letters,* Vol. 17, No. 24, November 26, 1981, pp. 901–902.
7. Microwave IMPATT Diodes Data Sheet (ND 8 Si CW Series), California Eastern Laboratories (NEC Corporation).
8. Misawa, T., and N.D. Kenyon, "An Oscillator Circuit with Cap Structures for Millimeter-Wave IMPATT Diodes," *IEEE Trans. Microwave Theory Tech.,* Vol. MTT-18, No. 11, November 1970, pp. 969–970.
9. Groves, I.S., and D.E. Lewis, "Resonant-Cap Structures for IMPATT Diodes," *Electronics Letters,* Vol. 8, No. 4, February 24, 1972, pp. 98–99.

10. Kiehl, R.A., "Novel Optical Control Techniques for Solid-State Radar Transmitters," *IEEE Trans. Microwave Theory Tech.,* Vol. MTT-28, No. 4, April 1980, pp. 409–413.
11. Fong, T.T., and H.J. Kuno, "Millimeter-Wave Pulsed IMPATT Sources," *IEEE Trans. Microwave Theory Tech.,* Vol. MTT-27, No. 5, May 1979, pp. 492–499.
12. Schweighart, A., H.P. Vyas, J.M. Borrego, and R. J. Gutmann, "Avalanche Diode Structures Suitable for Microwave-Optical Interactions," *Solid-State Electronics,* Vol. 21, No. 9, September 1978, pp. 1119–1121.
13. Vyas, H.P., R.J. Gutmann and J.M. Borrego, "The Effect of Hole versus Electron Photocurrent on Microwave-Optical Interactions in IMPATT Oscillators," *IEEE Trans. Electron Devices,* Vol. ED-26, No. 3, March 1979, pp. 232–234.
14. Seeds, A.J. and J.R. Forrest, "Reduction of FM Noise in IMPATT Oscillators by Optical Illumination," *Electronics Letters,* Vol. 17, No. 23, November 12, 1981, pp. 865–866.
15. Pitner, P.M., R.J. Gutmann, and J.M. Borrego, "Photocurrent Effects on Noise in Silicon IMPATT Oscillators," *IEE Proc.,* Vol. 129, Pt. I, No. 4, August 1982, pp. 149–152.
16. Maeda, M., S. Takahashi and H. Kodera, "CW Oscillation Characteristics of GaAs Schottky-Barrier Gate Field-Effect Transistor," *Proc. IEEE (Letters),* Vol. 63, No. 2, February 1975, pp. 320–321.
17. Johnson, K.M., "Large Signal GaAs MESFET Oscillator Design," *IEEE Trans. Microwave Theory Tech.,* Vol. MTT-27, No. 3, March 1979, pp. 217–227.
18. Kotzebue, K.L., and W.J. Parrish, "The Use of Large Signal S-Parameters in Microwave Oscillator Design," *Proc. IEEE Int. Symp. on Circuits and Systems,* April 21–23, 1975.
19. Sze, S.M. *Physics of Semiconductor Devices,* 2nd Ed., New York, John Wiley and Sons, 1981, Chapter 10.

Chapter 10
Optoelectronic Injection-Locking and Tuning of Oscillators

10.1 INTRODUCTION

In phased array antenna systems and communication systems, there is frequently a need to phase-lock several remotely located microwave sources. This is generally accomplished in existing systems by dividing and distributing the output of a single microwave source, or by electrically injecting a microwave reference signal to each of the oscillators. The first technique is inefficient because a large part of the reference signal is attenuated in the power divider as well as interconnecting transmission lines. The second technique, which is known as *electrical injection-locking,* avoids these power losses. However, in certain applications such as airborne phased array radar systems and satellite communication systems, the size and weight of the waveguides or coaxial cables, required to convey the reference signals, are still objectionable. To address this problem, an optical fiber link with transmitters employing semiconductor laser diodes capable of being directly modulated at microwave frequencies and receivers with photodiodes capable of detecting these signals have been proposed [1]. The fiber optic links offer several advantages. First, there is inherent isolation between the locked oscillator and the locking source. Second, optical fibers replace metal waveguides and coaxial cables, thereby reducing weight and volume. Third, microwave phase shifting can be easily introduced by delaying the optical signal. Finally, optical fiber offers immunity to electromagnetic interference and *electromagnetic pulse* (EMP).

In Chapter 9, we described optoelectronic switching and modulation of microwave oscillators. In this chapter, optoelectronic injection-locking and tuning of microwave oscillators will be discussed. There are two techniques for optoelectronically injection-locking an oscillator. The first technique is known as *direct optical injection-locking*. In this technique, the oscillator is injection-locked by focusing the light from an intensity-modulated laser diode onto the active area of the device [2–6]. As opposed to this scheme, we may first detect the intensity-modulated optical signal and electrically injection-lock the oscillator [7,8]. This technique is known as *indirect optical injection-locking*. The oscillator circuits considered in this chapter are based on IMPATT diodes and MESFETs. Finally, unmodulated laser diode illumination is used to tune an IMPATT or MESFET oscillator. The basic mechanism responsible for the change in the oscillator frequency is the same as that presented in sections 9.2.2 and 9.10 of Chapter 9.

10.2 BRIEF REVIEW OF CONVENTIONAL ELECTRICAL INJECTION-LOCKING TECHNIQUES AND EXPERIMENTS

10.2.1 CW Oscillators

Injection-locking is a technique used to stabilize the frequency and RF phase of a free-running oscillator. If a signal at a frequency, f_1, which is close to the free-running oscillator frequency, f_0, is injected into the oscillator, the oscillator frequency is pulled from the free-running state and synchronized or locked to that of the injected signal [9]. Thus, the frequency and RF phase of the oscillator are tied to those of the injected CW signal.

An injection-locked oscillator is configured as shown in Figure 10.1. In this setup, the locking signal is fed to the oscillator by means of a circulator, which provides isolation between the locking source and oscillator as well as the load and oscillator.

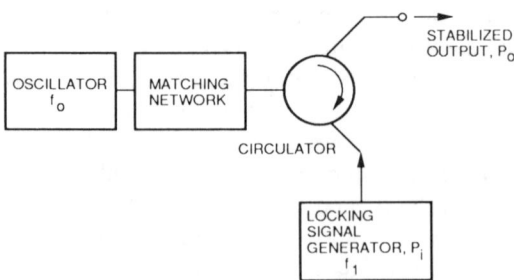

Figure 10.1 Schematic illustrating injection-locking technique of a CW oscillator.

An important characteristic of the injection-locked oscillator is the variation of the *locking range* with the *locking gain*. The locking range or *locking bandwidth* is defined as the frequency range across which the oscillator frequency remains locked to the frequency of the injected signal. The locking gain is defined as the ratio of the oscillator output power, P_0, and the injected signal power, P_i. A typical plot of the experimentally measured locking range *versus* the locking gain of an IMPATT oscillator [10] is shown in Figure 10.2. This curve has a characteristic slope of 20 dB/decade and is independent of whether the oscillator is an avalanche diode oscillator [11], transferred electron device oscillator [12], or a MESFET oscillator [13]. The second important attribute of injection-locked oscillators is the FM noise. Within the locking-range, the FM noise assumes the characteristics of the reference oscillator [9]. However, outside the locking-range, the noise is the same as that of the free-running oscillator. The AM noise is not affected by injection locking; it remains the same as that in the free-running case.

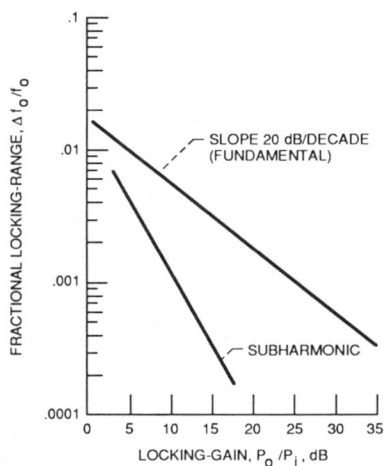

Figure 10.2 Locking-range *versus* locking-gain characteristic.

The frequency ratio is defined as the ratio of the oscillator frequency, and the locking signal frequency, which in Figure 10.2 is 1. In general, as the frequency ratio increases or when the oscillator is injection locked by a signal which is a subharmonic of the oscillator frequency the slope of the curve shown in Figure 10.2 becomes steeper [11]. That is, for a fixed locking-gain the locking-range decreases as the frequency ratio increases. The FM noise within the locking-range, in the case of a subharmonically injection-locked oscillator, degrades by a $20 \log N$ dB. Where N is the subharmonic number. As an example, if a reference oscillator, which is used to injection lock a free running oscillator through its fourth subharmonic, has a FM noise of -120 dBc/Hz then the FM noise of the locked oscillator will be at best -108 dBc/Hz. Finally, Lissajous pattern measurements show that as the oscillator frequency is pulled from one end of the locking range to the other, the RF phase of the oscillator changes from $-90°$ to $+90°$ [11].

10.2.2 Pulsed Oscillators

The object here is to injection lock a pulsed oscillator in order to generate RF pulses that are pulse-to-pulse coherent. That is, each RF pulse starts with the same phase as the preceding one. In order to achieve this, a CW locking signal at a frequency, f_1 which is separated by an amount of $+\Delta f$ from the frequency, f_0 of the oscillator, is injected into the pulsed oscillator setup shown in Figure 10.3. Phase locking takes place when the CW signal amplitude exceeds a critical value, typically -10 dBc [14].

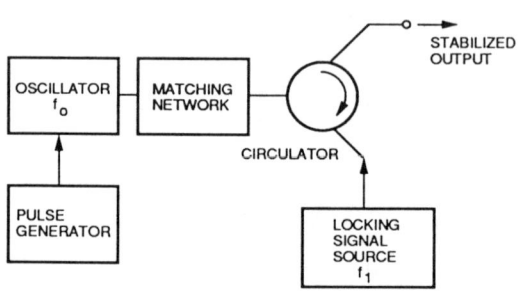

Figure 10.3 Schematic illustrating injection-locking technique of a pulsed oscillator.

In a phase-primed oscillator the CW locking signal power is insufficient to ensure phase locking, but is sufficient to be the dominant factor in the build up of oscillations from small to large signal. Thus, the locking signal impresses its own phase as an initial starting phase upon each pulse of the oscillator. The pulsed oscillator then deviates in phase continuously once large signal operations take place [14,15]. Typical powers required for phase priming are much less than for phase locking (typically -50 dBc), and may be at offset frequency of up to 1 GHz [14] at Ku band frequencies.

In an ideal pulsed oscillator the frequency spectrum consists of a series of sharp lines, referred to as Fourier lines, which are spaced apart by the PRF and with $(\sin x)^2/x$ amplitude. For this ideal case to be achieved, there must be perfect coherence between the starting phase of all the pulses of RF energy. In practice this will not be achieved, and there will be a degradation of the observed spectrum. If the degree of pulse-to-pulse phase incoherence has a small Gaussian error σ_ϕ, each Fourier line is reduced in magnitude by an amount of σ_ϕ^2 and this power loss can be observed as a white noise contribution between the Fourier lines. The ratio of the peak of the Fourier line to this noise is called the *peak-to-valley ratio*. Thus, a measure of the peak-to-valley ratio can be used to determine the pulse-to-pulse phase error σ_ϕ. A phase bridge setup for this measurement is described in [16]. Figure 10.4(a) shows

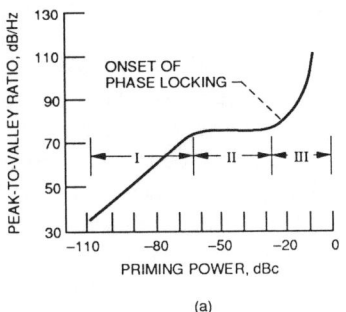
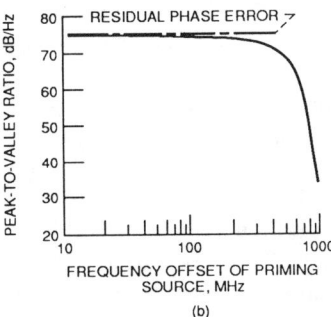

Figure 10.4 (a) Peak-to-valley ratio *versus* priming power at a fixed offset frequency; (b) peak-to-valley ratio *versus* frequency offset of priming source at a fixed priming power.

a typical plot of the peak-to-valley ratio for both phase-priming and phase-locking modes at a fixed offset frequency [14]. At low priming powers, the peak-to-valley ratio increases as the priming power is increased. This demonstrates that the injected signal is gaining control of the starting phase of each pulse. The flat portion of the curve corresponds to a residual phase error as the oscillator makes a transition from small-signal to large-signal conditions. For large priming powers, the interline noise assumes the characteristics of the injecting oscillator.

As the offset frequency changes, the coherency of the oscillator will also change. Figure 10.4(b) shows a typical plot of the peak-to-valley ratio at a fixed priming power [14]. The fall off at high offset frequency arises from the quality factor (Q) of the RF circuit. The flat portion of the curve corresponds to a constant residual phase error.

An important parameter associated with injection-locked, pulsed IMPATT oscillators is the amount of temperature-induced frequency chirp that can be tolerated [17]. Because the injection-locking bandwidth for a given amount of injection power is limited, the frequency chirping during the pulsing of the diode is to be minimized. A typical acceptable value for the frequency chirping is about 100 MHz over a pulse-width of 100 ns [17]. Techniques to minimize frequency chirping in pulsed IMPATT oscillators is explained in Section 9.4.1 of Chapter 9.

10.3 DESIGN OF A MILLIMETER-WAVE MICROSTRIP IMPATT DIODE OSCILLATOR

In Section 9.3 of Chapter 9, the designs of waveguide IMPATT diode oscillators were presented. These waveguide oscillators are capable of delivering several hundred milliwatts of CW power at millimeter-wave frequencies. In this section, we present the design of an IMPATT diode oscillator with microstrip transmission medium.

These oscillators are suitable for low-power applications, typically a few tens of milliwatts at millimeter-wave frequencies.

Let us suppose that there is a requirement for a microstrip IMPATT diode oscillator at 100 GHz. (This example is based on a design from [18].) The choice of the substrate material for the microstrip depends on several factors, but those which are important to this example are relative permittivity and mechanical strength. By choosing a substrate with low relative permittivity, the effective dielectric constant or guide wavelength will be closer to that in free space and circuit dimensions consequently will be larger. This results in ease of fabrication by relaxing the dimensional tolerances, thereby reducing the cost. Mechanical strength is also important to prevent breakage while handling. These considerations result in a choice of a soft substrate material such as RT/Duroid 5880 (Roger Corporation) with a relative permittivity of 2.2. RT/Duroid is available with a thick metal laminated to one side and copper foil on the other [19]. In addition to providing mechanical support, the thick metal facilitates heat sinking of active devices such as IMPATT, and will be discussed subsequently.

The basic oscillator circuit is shown in Figure 10.5. In this circuit, the dc bias network is designed as a low-pass filter. The low-pass filter allows the dc bias to reach the IMPATT diode, but prevents the RF power from leaking to the bias power supply. A typical low-pass filter for dc bias would consist of a two section high-low impedance line [20]. In this example, the low-impedance line is chosen as 20 Ω, and the high-impedance line is chosen as 170 Ω. The length of each section is approximately a quarter-wavelength.

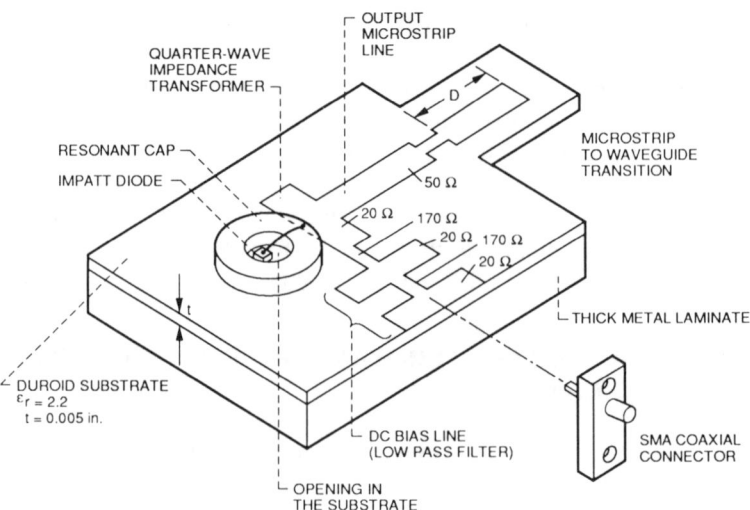

Figure 10.5 The circuit layout of a millimeter-wave microstrip IMPATT diode oscillator.

Commercially available IMPATT diodes are usually mounted in hermetically sealed packages with a copper heat sink [21]; for the oscillator circuit, the heat sink is embedded in the thick metal through a circular opening cut in the soft substrate.

In analogy to the waveguide resonant cap structure discussed in Section 9.3.2, a resonant cap is placed on top of the IMPATT diode. As has been observed, experimentally by varying the cap diameter, the oscillator frequency and power output can be fine tuned. The cap diameter derived through this process is 1.5 mm.

The IMPATT diode impedance is typically on the order of a few ohms. Hence, an impedance-matching transformer is required to match the diode impedance to the 50 Ω output microstrip line. In this example, a single-section, quarter-wave, impedance-matching transformer is used. The impedance of the quarter-wave section is 20 Ω.

At higher millimeter-wave frequencies, due to the inavailability of coaxial connectors, coupling the output of the oscillator to a rectangular waveguide is desirable. Hence, a microstrip-to-waveguide transition must be developed. The design of one such transition is discussed below and is based on Shih *et al.* [22]. The transition is schematically illustrated in Figure 10.6. At the transition, RF power is coupled

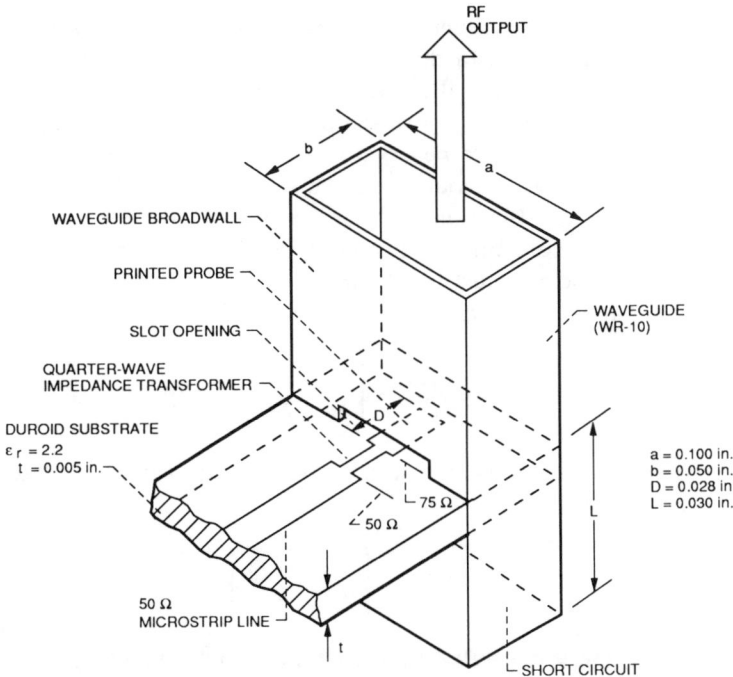

Figure 10.6 Schematic illustrating a microstrip-to-waveguide printed probe transition.

from the 50 Ω microstrip line to the waveguide by a printed probe. A printed probe is formed when a microstrip physically extends a distance D (Figure 10.6) through a narrow transverse slot cut in the broadwall of the waveguide. A backshort is placed at a longitudinal distance L from the probe to tune out the probe reactance. Thus, the input impedance of the probe is purely resistive, and is typically found to be about 75 Ω. Hence, the 75 Ω probe input impedance is matched to the 50 Ω output microstrip line by a quarter-wave impedance transformer.

Finally, as reported in [18], the oscillator was constructed and oscillated at a frequency of 97.3 GHz, with output power of 10 mW at a bias voltage and current of 20.8 V and 80 mA, respectively.

10.4 DESIGN OF A MICROSTRIP GaAs MESFET OSCILLATOR

In the case of two-terminal negative-resistance devices such as IMPATT diodes, the act of applying the proper dc bias is sufficient to generate negative resistance across the device terminals. However, in the case of MESFETs, the negative resistance that leads to oscillation must be induced by the choice of proper feedback elements and terminating impedances.

In general, a MESFET oscillator can be constructed by either series or parallel feedback circuit elements [23]. In this example, an oscillator with a source series feedback [24] is considered, and is illustrated in Figure 10.7(a). For the circuit shown in the figure to oscillate, the real part of the output impedance Re (Z_{out}), which is a function of both X_1 and X_3, should be negative [23]. Further, from maximum output power considerations, Re (Z_{out}) should be as large as possible [24]. Hence, the first step is to determine the combination of X_1 and X_3 that yield the maximum absolute value of Re (Z_{out}). To achieve this, an expression for Z_{out} in terms of the device and circuit impedance parameters must be obtained.

Let the small-signal S-parameters of the MESFET be represented as

$$[S] = \begin{bmatrix} S_{11} & S_{12} \\ S_{21} & S_{22} \end{bmatrix} \tag{10.1}$$

Next, this S-parameter matrix is transformed into an equivalent impedance matrix $[Z_a]$, which is represented as

$$[Z_a] = \begin{bmatrix} Z_{11a} & Z_{12a} \\ Z_{21a} & Z_{22a} \end{bmatrix} \tag{10.2}$$

Further, let the impedance matrix $[Z_b]$ for the series feedback impedance jX_3 be represented as

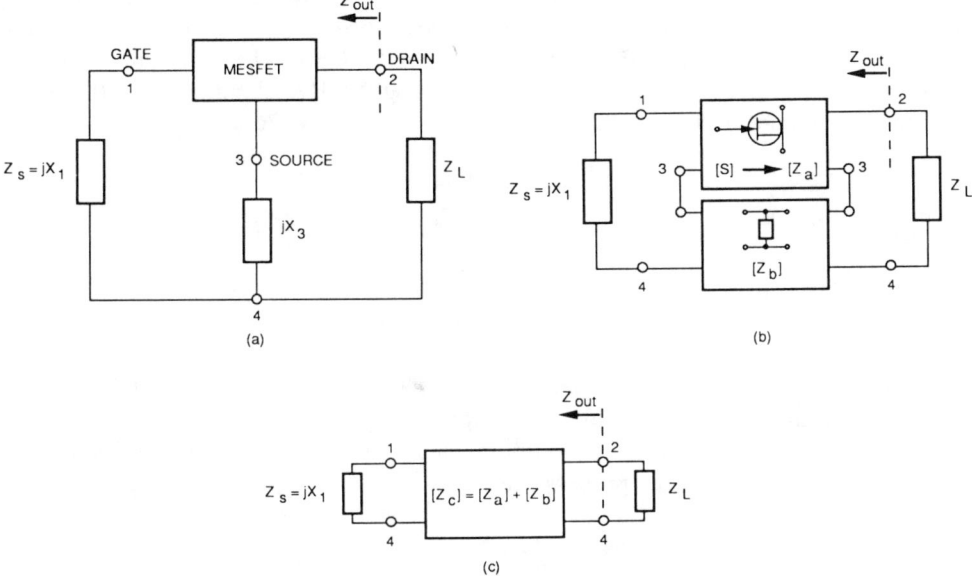

Figure 10.7 (a) Simplified equivalent circuit of a MESFET oscillator; (b) equivalent impedance matrix representation of the MESFET and the feedback element; (c) overall equivalent impedance matrix representation of the oscillator circuit.

$$[Z_b] = \begin{bmatrix} Z_{11b} & Z_{12b} \\ Z_{21b} & Z_{22b} \end{bmatrix} \tag{10.3}$$

By referring to Figure 10.7(b), the overall impedance matrix $[Z_c]$ for the two networks in series can be written as

$$[Z_c] = [Z_a] + [Z_b] \tag{10.4}$$

$$= \begin{bmatrix} Z_{11a} + Z_{11b} & Z_{12a} + Z_{12b} \\ Z_{21a} + Z_{21b} & Z_{22a} + Z_{22b} \end{bmatrix} \tag{10.5}$$

$$= \begin{bmatrix} Z_{11c} & Z_{12c} \\ Z_{21c} & Z_{22c} \end{bmatrix} \tag{10.6}$$

The output impedance Z_{out} is then obtained from the expression [25]:

$$Z_{out} = Z_{22c} - \left[\frac{Z_{21c} Z_{12c}}{Z_{11c} + Z_s} \right] \tag{10.7}$$

Let us suppose that the requirement is for an oscillator at 9 GHz. Further, let us suppose that the active device selected is a low noise GaAs FET (NE202), which has a maximum frequency of oscillation, f_{max}, of 50 GHz. The common-source S-parameters (magnitude and angle) of this device at 9 GHz, obtained from the manufacturer's data sheet at $V_{ds} = 2.0$ V and $I_{ds} = 20$ mA, is [26]:

$$S_{11} = 0.894, -71°$$
$$S_{21} = 3.414, 116°$$
$$S_{12} = 0.077, 45°$$
$$S_{22} = 0.603, -39°$$

The stability factor, k and the *maximum available gain* (MAG) determined from these S-parameters are 0.36 and 16.4 dB, respectively.

By resorting to (10.7), the optimum values of X_1 and X_3 for maximum Z_{out} obtained through an iterative process are

$$X_1 = -25 \ \Omega$$
$$X_3 = 50 \ \Omega$$

and the corresponding Z_{out} is

$$Z_{out} = -7.805 - j70.353 \ \Omega$$

The next step is to obtain the load impedance Z_L. The imaginary part of the load impedance $\text{Im}(Z_L)$ is directly determined by the resonance condition at the desired frequency of oscillation:

$$\text{Im}(Z_L) = -\text{Im}(Z_{out}) \qquad (10.8)$$

However, the real part $\text{Re}(Z_L)$ is determined by the equation [24]:

$$\text{Re}(Z_L) = 1/3 \ \text{Re}(Z_{out}) \qquad (10.9)$$

Hence, the load impedance Z_L is

$$Z_L = 2.60 + j70.35 \ \Omega \qquad (10.10)$$

The load impedance is realized with an arbitrary load resistance R_0, series capacitance C, and microstrip line having a characteristic impedance Z_{02} and an electrical length of Θ_2, as shown in Figure 10.8. The load resistance R_0 is realized by

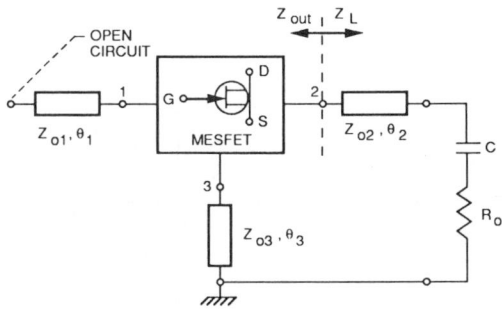

Figure 10.8 Practical FET oscillator with microstrip elements.

a 50 Ω chip resistor, and the capacitance C is realized by a 0.1 pF chip capacitor. The reactance X_1 is realized with a microstrip line having a characteristic impedance and electrical length of Z_{01} and Θ_1, respectively, and terminating in an open circuit. Similarly, the reactance X_3 is realized with a microstrip line having a characteristic impedance and electrical length of Z_{03} and Θ_3, respectively, and terminating in a short circuit. The characteristic impedances can take any arbitrary value, but they are normally chosen as 50 Ω for ease of impedance matching. The electrical lengths are read from the Smith chart. The microstrip characteristic impedances and electrical lengths determined for the above circuit are as follows:

$$Z_{01} = 50 \ \Omega, \ \Theta_1 = 64°$$
$$Z_{02} = 50 \ \Omega, \ \Theta_2 = 129°$$
$$Z_{03} = 35 \ \Omega, \ \Theta_3 = 55°$$

Finally, the entire circuit can be fabricated on a standard 0.025-inch thick alumina substrate.

10.5 DIRECT OPTICAL INJECTION-LOCKING AND TUNING OF CW OSCILLATORS

10.5.1 Principle of Operation

Direct optical injection-locking of a microwave oscillator is schematically illustrated in Figure 10.9. In this setup, the locking signal or the reference signal at a frequency, f_1, intensity-modulates the light from a semiconductor laser diode. The intensity-modulated optical signal from the laser then illuminates the active region of the IMPATT or the MESFET in the oscillator circuit and is absorbed. In this manner, the signal

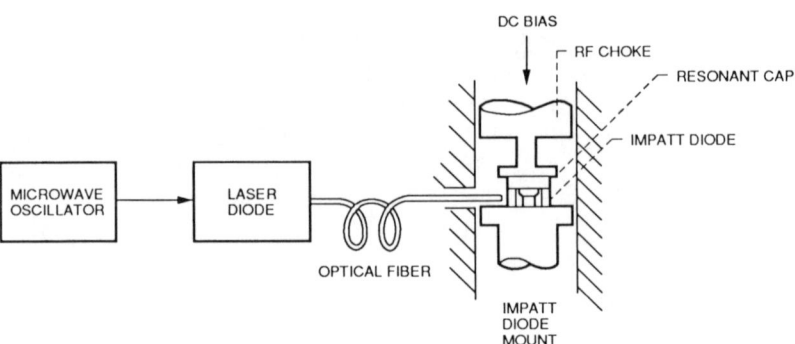

Figure 10.9 Schematic illustrating direct optical injection-locking technique of a CW IMPATT oscillator.

carried by the light is coupled or injected into the oscillator circuit. If f_1 is equal to the oscillator frequency f_0, fundamental frequency injection-locking occurs.

The limitations of this technique are the poor coupling efficiency between the laser beam and the narrow device active region, and the maximum modulation frequency of commercially available semiconductor diode lasers is limited to a few GHz [27]. The latter limitation can be circumvented to some extent by resorting to subharmonic injection-locking, or by optically generating an injection-locking signal [28].

In subharmonic injection-locking, f_1 is a subharmonic of f_0. Thus, the subharmonic technique allows injection-locking of oscillators operating at frequencies much higher than the laser diode modulation frequency. However, as in the case of conventional injection-locking, the locking range or bandwidth decreases as the frequency ratio increases [2].

The technique of optically generating a microwave signal by using semiconductor laser diodes has been discussed in Chapter 7. The technique consists of simultaneously injection-locking the adjoining longitudinal modes of a slave laser by the FM sidebands of a master laser, which is directly modulated by a sinusoidal signal. The two longitudinal modes of the slave laser are then made to illuminate and beat together in the active regions of the IMPATT or MESFET in the oscillator circuit, thereby generating the locking signal.

10.5.2 Experimental Demonstration of Direct Optical Injection-Locking and Tuning

Yen [3] and Seeds and Forrest [4] have demonstrated the feasibility of direct optical injection-locking of a silicon IMPATT oscillator. A laser diode intensity-modulated at a frequency of 2.704 GHz was used for illuminating the IMPATT. The emitted optical illumination was directly focused on the IMPATT diode active region. Because the free-running frequency of the IMPATT oscillator is 8.133 GHz, injection-locking occurred at the third subharmonic.

Seeds *et al.* [29] have also demonstrated frequency tuning of oscillators by using an unmodulated laser diode to illuminate an IMPATT device, and then producing a frequency shift. The shift in the oscillator frequency is found to be proportional to the optically generated current. The tuning slope calculated from the measured data is typically in the range of few hundred MHz/mA. This is higher than the case of conventional electronic tuning of IMPATT oscillators for which the bias tuning factor is about 1 MHz/mA [21]. The tuning speed of optically tuned IMPATT oscillators is on the order of 70 MHz/μs [29], which is smaller by a factor of two than desirable for a local oscillator in a frequency agile system [30].

Injection-locking of GaAs MESFET oscillators has been reported in [5,6]. A laser diode intensity-modulated at a frequency of 2.8 GHz was used to illuminate the MESFET. The emitted optical illumination was directly focused on the active region of the MESFET. Because the free-running frequency of the MESFET oscillator is 2.8 GHz, injection-locking occurred at the fundamental frequency. Injection-locking of a GaAs MESFET oscillator by an optically generated microwave signal has also been demonstrated. The details of the experiment are reported in [31].

Finally, experiments on optical tuning of GaAs MESFET oscillators yielded a tuning factor on the order of a few MHz/mA [32]. The tuning speed is expected to be slow when compared to an IMPATT, due to the trapping of carriers as explained in Section 9.10 of Chapter 9.

10.6 INDIRECT OPTICAL INJECTION-LOCKING OF CW OSCILLATORS

10.6.1 Principle of Operation

Indirect optical injection-locking of a microwave oscillator is schematically illustrated in Figure 10.10, where the locking signal or the reference signal at a fre-

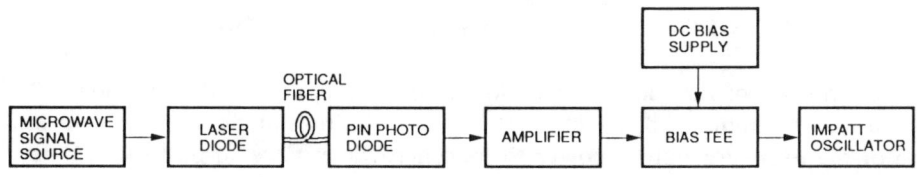

Figure 10.10 Schematic illustrating indirect optical injection-locking technique of a CW IMPATT oscillator.

quency, f_1 intensity-modulates a semiconductor laser diode operating at a frequency f_c. The intensity-modulated optical signal is then detected by a photodiode. The detected output is filtered, amplified, and then electrically injected into the oscillator. This method has the advantage of more efficient coupling of the laser light to the large-area photodiode than is the case for the thin active region of the IMPATT diode as in direct locking. However, the disadvantages of this technique is that it requires additional components, which may add to the cost and complexity of the circuit.

10.6.2 Experimental Demonstration of Indirect Optical Injection-Locking

To demonstrate the feasibility of indirect injection-locking, Herczfeld et al. [7] used a silicon IMPATT oscillator. A laser diode intensity-modulated at a frequency of 3.235 GHz was used in this experiment. This modulation frequency happens to be close to the laser relaxation frequency so that the nonlinearity in the laser diode characteristic caused a large number of harmonics of the modulating signal, as explained in Section 8.5.1 of Chapter 8. These harmonics also intensity-modulate the light output from the laser diode. The intensity-modulated light output from the laser was then detected by a photodiode. The detected output hence contained the original signal at 3.235 GHz as well as its harmonics. The fourth harmonic was selected by filtering, amplified, and electrically injected through a bias tee to the biasing port of the free-running IMPATT oscillator. Injection-locking was observed at three times the frequency of the injected electrical signal (i.e., at 38.820 GHz). Thus, with respect to the master oscillator, the injection-locking process occurred at the twelfth harmonic.

Indirect injection-locking of a GaAs MESFET oscillator has been reported [8]. A laser diode intensity-modulated at a frequency of 3.593 GHz was used in this experiment. The emitted light was first detected by a photodiode, and the optically generated third harmonic was selected as in the previous example. The filtered signal was amplified and injected into a free-running GaAs MESFET oscillator. Because the free-running frequency of the oscillator was 21.5 GHz, injection-locking occurred at the sixth harmonic with respect to master oscillator.

10.7 COMPARISON OF DIRECT AND INDIRECT OPTICALLY INJECTION-LOCKED CW OSCILLATOR PERFORMANCE

The performance of optically injection-locked oscillators described in Sections 10.5.2 and 10.6.2 is summarized in Table 10.1. The definition of the locking gain is the ratio of the oscillator output power to the microwave power used to modulate the semiconductor laser diode [3]. Table 10.1 also presents typical performance parameters of conventional electrically injection-locked IMPATT and MESFET oscillators for comparison.

Table 10.1
Comparison of Direct and Indirect Optical Injection-Locking Characteristics of Impatt and MESFET CW Oscillators

	IMPATT				MESFET			
	Conventional	Optical			Conventional	Optical		
		Direct	Indirect			Direct	Indirect	
								Conventional
Frequency Ratio	1:1	3:1	12:1		1:1	1:1	4:1	1:1
Locking Range (MHz)	650	1	2 (Amplifier gain = 22 dB) 132 (Amplifier gain = 45 dB)			5	18 (Amplifier gain = 20 dB) 84 (Amplifier gain = 50 dB)	200
Approximate Locking Gain (dB)	18	−14	—			−21	—	20
Phase Noise	−90 dBc/Hz at 10 kHz offset carrier	—	−55 dBc/Hz at 5 kHz offset carrier			−70 to −80 dBc/Hz at 10 kHz offset carrier	−57 dBc/Hz at 1 kHz offset carrier	−80 to −90 dBc/Hz at 10 kHz offset carrier
Reference	[21]	[4]	[7]			[5,6]	[8]	[13]

The indirect optical injection-locking technique is observed to yield a larger locking range than direct optical injection-locking techniques. This is because the locking signal is amplified before injection. However, the penalty is a degradation of the phase noise of the injection-locked oscillator.

For identical frequency ratios, direct optical injection-locking yields a much lower locking range than conventional electrical injection-locking techniques. This is because of the low coupling efficiency between the optical fiber and the device's active area. One technique to improve coupling efficiency would be to use an IMPATT diode with a specially etched optical window for illumination, as explained in Section 9.5 of Chapter 9.

Direct injection-locking techniques would be impractical in the case of pulsed IMPATT oscillators. This is because the locking-bandwidth, in the case of direct optical injection-locking, is small when compared to the chirp bandwidth of the pulsed oscillator. Hence, for pulsed oscillators, indirect optical injection-locking is preferable.

Finally, the tuning factor of optically tuned IMPATT oscillators, as compared to conventional bias tuned oscillators, is found to be superior. However, the tuning speed is observed to be slower by a factor of two. The lower tuning speed may be due to the trapping of carriers that occur in the semiconductor substrate in a manner similar to that of an optically illuminated MESFET, which is explained in Section 9.10 of Chapter 9.

REFERENCES

1. Sobol, H., "The Application of Microwave Techniques in Lightwave Systems," *IEEE J. Lightwave Technol.*, Vol. LT-5, No. 3, March 1987, pp. 293–299.
2. Yen H.W., and M.K. Barnoski, "Optical Injection Locking and Switching of Transistor Oscillators," *Applied Physics Letters*, Vol. 32, No. 3, February 1978, pp. 182–184.
3. Yen, H.W., "Optical Injection Locking of Silicon IMPATT Oscillators," *Applied Physics Letters*, Vol. 36, No. 8, April 1980, pp. 680–683.
4. Seeds, A.J., and J.R. Forrest, "Initial Observation of Optical Injection Locking of an X-Band IMPATT Oscillator," *Electronics Letters*, Vol. 14, No. 25, December 1978, pp. 829–830.
5. DeSalles, A.A.A., J.R. Forrest, "Initial Observations of Optical Injection Locking of GaAs Metal Semiconductor Field Effect Transistor Oscillators," *Applied Physics Letters*, Vol. 38, No. 5, March 1981, pp. 392–394.
6. DeSalles, A.A.A., "Optical Control of GaAs MESFET's," *IEEE Trans. Microwave Theory Tech.*, Vol. MTT-31, No. 10, October 1983, pp. 812–820.
7. Herczfeld, P.R., A.S. Daryoush, A. Rosen, A.K. Sharma, and V.M. Contarino, "Indirect Subharmonic Optical Injection Locking of a Millimeter-Wave IMPATT Oscillator," *IEEE Trans. Microwave Theory Tech.*, Vol. MTT-34, No. 12, December 1986, pp. 1371–1375.
8. Daryoush, A.S., P.R. Herczfeld, R. Glatz, and A.P.S. Khanna, "Phase and Frequency Coherency of Multiple Optically Synchronized 20 GHz FET Oscillators for Satellite Communications," *IEEE MTT-S Int. Microwave Symp. Digest*, 1987, pp. 823–826.
9. Kurokawa, K., "Injection Locking of Microwave Solid-State Oscillators," *Proc. IEEE*, Vol. 61, No. 10, October 1973, pp. 1386–1410.

10. Shaw, R.C., and H.L. Stover, "Phased-Locked Avalanche Diode Oscillators," *Proc. IEEE (Letters)*, Vol. 54, No. 4, April 1966, pp. 710–711.
11. Chien, C.H., and G.C. Dalman, "Subharmonically Injected Phase-Locked IMPATT-Oscillator Experiments," *Electronics Letters*, Vol. 6, No. 8, April 1970, pp. 240–241.
12. Oltman, H.G., and C.H. Nonnemaker, "Subharmonically Injection Phase-Locked Gunn Oscillator Experiments," *IEEE Trans. Microwave Theory Tech.*, Vol. MTT-17, No. 9, September 1969, pp. 728–729.
13. Tajima, Y., and K. Mishima, "Transmission-Type Injection Locking of GaAs Schottky-Barrier FET Oscillators," *IEEE Trans. Microwave Theory Tech.*, Vol. MTT-27, No. 5, May 1979, pp. 386–391.
14. Brookbanks, D.M., and B.J. Buck, "The Switch on Characteristics and Noise of Pulsed Read IMPATTs in Ku Band," *IEEE MTT-S Int. Microwave Symp. Digest*, 1983, pp. 215–217.
15. Udelson, B.J., and R.E. Hines, "Effects of a CW Injected Signal on a Pulsed Avalanche Oscillator," *Proc. IEEE (Letters)*, Vol. 57, No. 11, November 1969, pp. 2091–2092.
16. Fong, T.T., and H.J. Kuno, "Millimeter-Wave Pulsed IMPATT Sources," *IEEE Trans. Microwave Theory Tech.*, Vol. MTT-27, No. 5, May 1979, pp. 492–499.
17. Yen, H.-C., and K. Chang, "A 63-W W-Band Injection-Locked Pulsed Solid-State Transmitter," *IEEE Trans. Microwave Theory Tech.*, Vol. MTT-29, No. 12, December 1981, pp. 1292–1297.
18. Morgan, G.B., "Microstrip IMPATT-Diode Oscillator for 100 GHz," *Electronics Letters*, Vol. 17, No. 16, August 1981, pp. 570–571.
19. Rogers Corporation, Thick-Metal Cladding on RT/Duroid Microwave Circuit Dielectric Laminates, Product Note RT 5.3.2.
20. Matthaei, G.L., L. Young, and E.M.T. Jones, *Microwave Filters, Impedance Matching Networks, and Coupling Structures*, Norwood, MA, Artech House, 1980, Section 7.03, p. 372.
21. Hughes Millimeter-Wave Products Catalogue (1987–88), Section 2, Power Generating Components, p. 46; Section 3, MM-Wave Diodes, p. 79; Section 7, Coherent Instrumentation Front End, p. 150.
22. Shih, Y.C., T.N. Ton, and L.Q. Bui, "Waveguide-to-Microstrip Transitions for Millimeter-Wave Applications," *IEEE MTT-S Int. Microwave Symp. Digest*, 1988, pp. 473–475.
23. Gonda, J., "Large Signal Transistor Oscillator Design," *IEEE MTT-S Int. Microwave Symp. Digest*, 1972, pp. 110–112.
24. Maeda, M., K. Kimura, and H. Kodera, "Design and Performance of X-Band Oscillators with GaAs Schottky-Gate Field-Effect Transistors," *IEEE Trans. Microwave Theory Tech.*, Vol. MTT-23, No. 8, August 1975, pp. 661–667.
25. Spence, R., *Linear Active Networks*, New York, Wiley-Interscience, 1970, Section 3.7, p. 96.
26. NEC Corporation (California Eastern Laboratories), Data Sheet on NE 202 Super Low Noise Hetero Junction FET.
27. High Speed Fiber-Pigtailed GaAlAs Laser Diodes and Transmitters (Specifications for TSL-Series Laser Transmitters), Ortel Corporation.
28. Goldberg, L., A.M. Yurek, H.F. Taylor, and J.F. Weller, "35 GHz Microwave Signal Generation with an Injection-Locked Laser Diode," *Electronics Letters*, Vol. 21, No. 18, August 1985, pp. 814–815.
29. Seeds, A.J., J.F. Singleton, S.P. Brunt, and J.R. Forrest, "The Optical Control of IMPATT Oscillators," *IEEE J. Lightwave Technol.*, Vol. LT-5, No. 3, March 1987, pp. 403–411.
30. Weissglas, P., S. Svensson, "Local Oscillators for Frequency Agile Systems," *Microwave J.*, Vol. 20, No. 1, January 1977, pp. 35–38.
31. Goldberg, L., C. Rauscher, J.F. Weller, and H.F. Taylor, "Optical Injection Locking of X-Band FET Oscillator Using Coherent Mixing of GaAlAs Lasers," *Electronics Letters*, Vol. 19, No. 20, September 1983, pp. 848–850.

32. Sun, H.J., R.J. Gutmann, and J.M. Borrego, "DC and Pulse-Light Illuminated Optical Responses of Microwave GaAs-MESFET Oscillators," *IEE Proc.,* Vol. 131, Pt. I, No. 1, February 1984, pp. 31–37.

INDEX

Absorption
    coefficient, 72
    depth, 158
    infrared, 145
    region, 76, 101
    ultraviolet, 145
Acceptance angle, 146
Acceptors, 89
Active channel, 106, 107
Admittance, 196
Aging loss, 151
Amplifier, 131, 133, 135–137
Annealing, 65
Antennas, 6–8
Antireflection coating, 79
Atomic Si sheet density, 111
Attenuation, 145
Avalanche
    current, 196–199
    delay, 198, 199
    multiplication, 101
    photodiode, 101
    photodiode (SAM), 101
    photodiode (SAGM), 101

Band gap energy, 28, 71
Bandwidth, 39, 40, 63, 88, 98, 139
Bandwidth distance product, 145
Barrier height, 90
Beam steering, 5
Bessel function, 168
Binomial series, 54
Bit rate, 106, 107
Boltzmann's constant, 129
Bragg grating, 42
Branching waveguide, 50

Breakdown voltage, 180
Broadening, pulse, 143
Buffer layer, 55, 56, 77, 78, 102
Buried
    channel waveguide, 48, 58
    heterostructure laser, 38–40
Butt coupling, 34

Cable television, 7
CAD (Computer aided design), 153
Capacitance, 38, 78, 93, 98, 100, 161
Carrier confinement, 39
    drift current density, 73
    injection, 39
    life-time, 84
    recombination, 15, 34, 86, 161
    trapping, 89
Catastrophic mirror damage, 37
Cellular telephone, 7
Characteristic impedance, 56
Chromatic (material) dispersion, 142–144
Cladding, 143, 144
Coaxial cable, 121, 122
Coherent communication, 2
Conduction band, 71
Confinement factor, 18, 19, 34
Conformal mapping technique, 56
Connector, 146, 148–151
    loss, 146
Core, 142–144
Coupler loss, 147
Coupling efficiency, 32, 58, 203
Coplanar
    stripline, 56–58
    stripline electrode, 56
    waveguide, 163

Critical angle, 50
 field, 196–198
Crosstalk, 3
Crystalline, 15
Current
 displacement, 71
 gain, 116
 generation-recombination, 99
 leakage, 191
 recombination, 99
 thermionic-emission, 98
 tunneling, 99
Cut-off wavelength, 72

Damage threshold, 66
Dark current, 90, 91, 98
Delay, 198, 199
Depletion region, 70, 71, 73, 96
Dielectric constant
 specific, 55
 tensor, 51
Differential quantum efficiency, 28, 40, 149
Diffusion
 coefficient, 72
 current, 73
 equation, 73
 length, 71, 191
Direct-band-gap semiconductors, 15
 modulation, 126
Dispersion, 142–145
 shifted optical fiber, 144
Distributed feedback lasers (InP-InGaAsP), 42
Donors, 89
Double heterostructure laser, 15–17
Drift current, 73
Drive voltage, 62, 63
Dynamic range, 142
 spurious free, 142

E^x_{pq} mode, 48
E^y_{pq} mode, 48
Electric field intensity, 49
Electrode
 attenuation, 58
 gap, 60
 impedance, 56
 length, 59, 62, 63
Electronic warfare, 6
Electro-optic
 coefficient, 52
 effect, 52

modulator, 47, 59
Electron-hole pair, 71, 84
 photocurrent, 205
Electronic warfare, 6
EMI (electromagnetic interference), 121
Emission
 spontaneous, 14, 15
 stimulated, 14, 15
EMP (electromagnetic pulse), 121
Energy band
 diagram, 17
 gap, 16
Enhancement of oscillations, 202
Epitaxial growth, 15
Equivalent circuit model, 78, 93, 94, 100, 101, 179–181
Etched optical window, 203–205
Evanescent
 fields, 147
Even mode, 22
Excess loss, 147
External modulation, 126
Extinction ratio, 60–62
Extraordinary optical index, 64

f_{max} (maximum frequency of oscillation), 116
f_t (unity current gain cut-off frequency), 1, 116
Fabry-Perot, 41, 43
Facet, 42
 reflectivity, 26, 27
Far field pattern, 30, 31
Feedback, optical, 129
Fermi level, 90
Ferroelectric crystal, 51
Fiber, 32–34, 58, 59, 121–125
Field effect transistor, 1, 3–5, 106
 dc characteristic, 111
 device structure, 107
 microwave characteristic, 114
 noise characteristic, 116
Finline, 163
Fourier lines, 216
Frequency
 stability (with temperature), 173
 modulation, 206, 211
 pulling, 174
 tuning, 225
 chirping, 203
Fresnel reflection, 72, 84
Friis formula, 137

Fundamental mode, 48

GaAlAs (gallium aluminum arsenide), 1, 38, 39
GaAs (gallium arsenide), 1, 2, 107–110
GaAs-GaAlAs laser, 38, 39, 43
Gain, 102, 103
GaInAsP (gallium indium arsenide phosphide), 1, (see also InGaAsP)
Gain bandwidth product, 87, 104
Gamma ray, 190
Gap
   series conductance, 160
   shunt conductance, 160
Gaussian
   error, 216
   index profile, 48–50
Gold film, 98
Grating, 42, 43
Gunn diode, 3
   oscillator, 173

Half-wave voltage, 60
Harmonic distortion, 139, 184
   generation, 226
HEMT (high electron mobility transistor), 106, 107, 109–111, 113–117
Heterojunction, 4, 15
   bipolar transistor, 4
Heterostructure, 15–17, 19
Hole photocurrent, 205
Hydroxyl impurity, 145

Impact ionization rate, 197
IMPATT diode, 196–206
   flip chip, 204, 205
   top mounted, 204, 205
Index ellipsoid, 51, 52
Index of refraction, 54, 55
InGaAs (Indium gallium arsenide), 110–111
InGaAsP (Indium gallium arsenide phosphide), 1, 40–43, 107
Injection locking, 170
   CW oscillators, 214
   direct, 223
   FM sideband, 168, 169
   indirect, 225
   pulsed oscillators, 216
InP (Indium phosphide), 1, 2
Insertion loss, 161
Instrumentation, 7, 8
Intensity, 28

$1/e$ point, 48, 58
   modulation, 121
   pattern, 20
Interdigital structure, 89, 90
Interferometry, 59
Intermodulation distortion, 139, 184, 185
   second order, 139
   third order, 139
Internal reflection, 20
Intrinsic layer, 70
Ionizing radiation, 189–191
Isolation, 161, 182
ITO (Indium tin oxide), 100

Johnson noise, 81, 82, 88, 185

Laser
   buried crescent (InP-InGaAsP), 40–42
   buried heterostructure (GaAs-GaAlAs), 38, 39
   buried heterostructure (InP-InGaAsP), 40, 41
   cavity, 19
   chirp, 47
   constricted mesa (InP-InGaAsP), 40–42
   current modulation, 32
   diode, 3–5, 32
   distributed feedback, 41, 42
   double heterostructure, 15–17
   index guided, 39
   intensity characteristic, 28
   optical spectra-free running, 29, 166, 167
   optical spectra-frequency modulation, 168
   slave, 169
   window, 37
Lattice constant, 76
   match, 76
   mismatch, 76
Leakage current, 78, 81
Lens
   conical, 33, 34
   hemispherical microlens, 33, 34
   spherical, 33, 34
   tapered fiber with high-index lens, 33, 34
Life-time
   effective, 84
   photon, 34, 37
   spontaneous recombination, 34
   surface, 84, 86
   volume, 84, 86
Light-induced voltage, 112, 113

LiNbO$_3$ (lithium niobate)
  Y-cut, 54, 55
  Z-cut, 54, 55, 59
Linewidth, 2
Links
  external modulated, 126
  directly modulated, 126
Locking gain, 214, 215
Locking range, 214, 215
Longitudinal mode, 166, 167
LPE (liquid phase epitaxy), 1

Mach-Zehnder
  interferometry, 47
  modulator, 8
MAG (maxium available gain), 116
Magnetic field
  stray, 174, 175
Material dispersion, 142–144
Maxwell's equations, 20
MBE (molecular beam epitaxy), 1
Mesa structure, 77, 78
MESFET (metal semiconductor field effect
    transistor), 77, 107, 195, 196
  equivalent circuit, 114–116
Microstrip
  line, 78, 79
  oscillator, 217–223
  switch, 157–159
Minority carriers, 190, 191
Misalignment loss, 151
Missile guidance, 7
MMIC (Monolithic microwave integrated
    circuit), 1
Mobility, 85
MOCVD (metallo-organic chemical vapor
    deposition), 89
Mode
  TE, 25, 26, 48
  TE, even, 22
  TE, odd, 25
  TM, 25, 48
Modulation depth, 88
  index, 88
Modulators
  intensity, 4
Monolithic integration, 1–3
MSM (metal-semiconductor-metal), 4, 90, 91
Multiplication
  factor, 180

region, 101
Multiquantum well, 4

n-type semiconductor, 16
Near field pattern, 30, 31
Neutron fluence, 190
Noise
  AM, 205
  equivalent power, 82, 83, 88, 89, 105
  FM, 205
  generation-recombination, 88
  Johnson, 88, 185
  shot, 88, 186
  temperature, 186
  thermal, 88, 185–188

Ohmic contact, 83, 95
OMMIC (optical microwave monolithic
    integrated circuit), 2
Optical
  anisotropy, 51
  damage, 65
  fiber, 32, 34
  waveguide, 48–50
Optoelectronic
  gating, 157
  microwave signal generation, 165
  oscillator injection-locking, 213
  oscillator modulation, 195
  oscillator switching, 195
  oscillator tuning, 213
  switching, 157, 158
  switch matrix, 178
Overlap integral, 60

Parasitic reactance, 93
Peak-to-valley ratio, 216, 217
Permeability, 20
Permittivity, 20
Phased-arrays, 5
Phase priming, 216, 217
Photobleaching, 189
Photocapacitance, 210
Photoconductive detector, 5, 83–87
Photoconductive effect, 210, 211
Photoconductive gain, 84
Photoconductivity (picosecond), 8
Photocurrent, 205
Photo diode, 69–71, 101
Photo responsivity, 91, 100
Photovoltaic effect, 210

Photon
    density, 33, 34
    lifetime, 34
Pigtail, 149
PIN photodiode, 4, 70
Planck's constant, 73
Pockels
    constant, 55
    effect, 52
Polyimide layer, 40, 41
Population inversion, 15
Propagation constant, 21
Pseudomorphic layer, 76, 109, 110
p-type semiconductor, 73
Pulling figure, 174
Push-pull operation, 59

Quantum efficiency, 80, 96, 98
    external, 95
    internal, 74, 97
Quantum noise, 81, 82
Quaternary compounds, 1
Quenching of oscillations, 202

Radar, 5, 6
Radiation hardness, 43, 66, 83, 189, 191
Rate equations, 33, 34
Rayleigh scattering, 145
Rectangular waveguide, 121
Reference frequency distribution, 5
Reflection coefficient, 126
Relative intensity noise, 127–129, 171–172
Relaxation oscillation, 37
    resonance, 36, 37
Reliability, 43, 83, 192
Remote location of antennas, 6
Repetition rate, 161
Response speed, 75, 98
Responsivity, 75, 102, 103
Reverse biased junction, 70, 71
Reverse saturation current, 197
RFI (radio frequency interference), 121
Richardson constant, 92, 99
Rise time, 87

Satellites, 177
Saturation velocity, 74
Scattering loss, 145
Schottky barrier, 90, 92, 93
Schottky barrier photodiode, 95–100
Schwartz-Christoffel transformation, 56

Second harmonic, 139
Semi-insulating substrate, 77
Series feedback oscillator, 206, 207
Shock, 175
Shot noise, 81, 82, 129, 186
Sidebands, 168
Signal generation, 169
Signal processing, 6
Signal-to-noise ratio, 80, 82, 88, 105, 142, 171, 185, 187–189
Single-mode, 58
Skin depth, 57, 58
Small-signal parameters, 206–208
Spectral
    line width, 2, 29, 30, 43
    response, 79
Splice loss, 147
Spontaneous emission, 14, 15
Stability factor, 206
Star coupler, 150
Stimulated emission, 14, 15
Subharmonic injection-locking, 215, 224
Surface
    resistance, 57
Switch
    crosspoint, 178
    matrix (optical), 178
    matrix (electronic), 183
Switching speed, 162

Tactical aircraft, 5
TE (transverse electric modes), 21, 22, 26, 27
Temperature
    coefficient of refractive indexes, 55
    stability, 173
Thermal noise, 88, 129
Thermionic-emission theory, 99
Thevenin equivalent circuit, 180
Thin-film waveguide, 48
Threshold
    current, 17
    optical damage, 66
TM (transverse magnetic modes), 25, 26, 27
Transit time, 88, 93, 198
    bandwidth-efficiency product, 74
    cut-off frequency, 74
Transmit-receiver module, 5
Traps for carriers, 89, 211
Traveling wave modulators, 59
Tuning, 224, 225

2 DEG (two-dimensional electron gas), 108–110

Valence band, 71
Velocity mismatch, 64
Vibrations, 175
Volume photoconductivity, 159
VPE (vapor phase epitaxy), 89

Wave equation, 23

Waveguide cavity
    reduced height, 199, 200
    standard height, 201
Waveguide-channel, 58
Waveguide mode
    asymmetric ratio, 58
    depth, 58
    eccentricity, 58
    width, 58